WHERE TO FIND GOLD IN

CALIFORNIA

AN EPIC GOLD JOURNAL OF CLASSIC INFORMATION

RESEARCHED, COMPILED AND WRITTEN

by

DELOS TOOLE

Chief Executive Producer

TOOLE ©

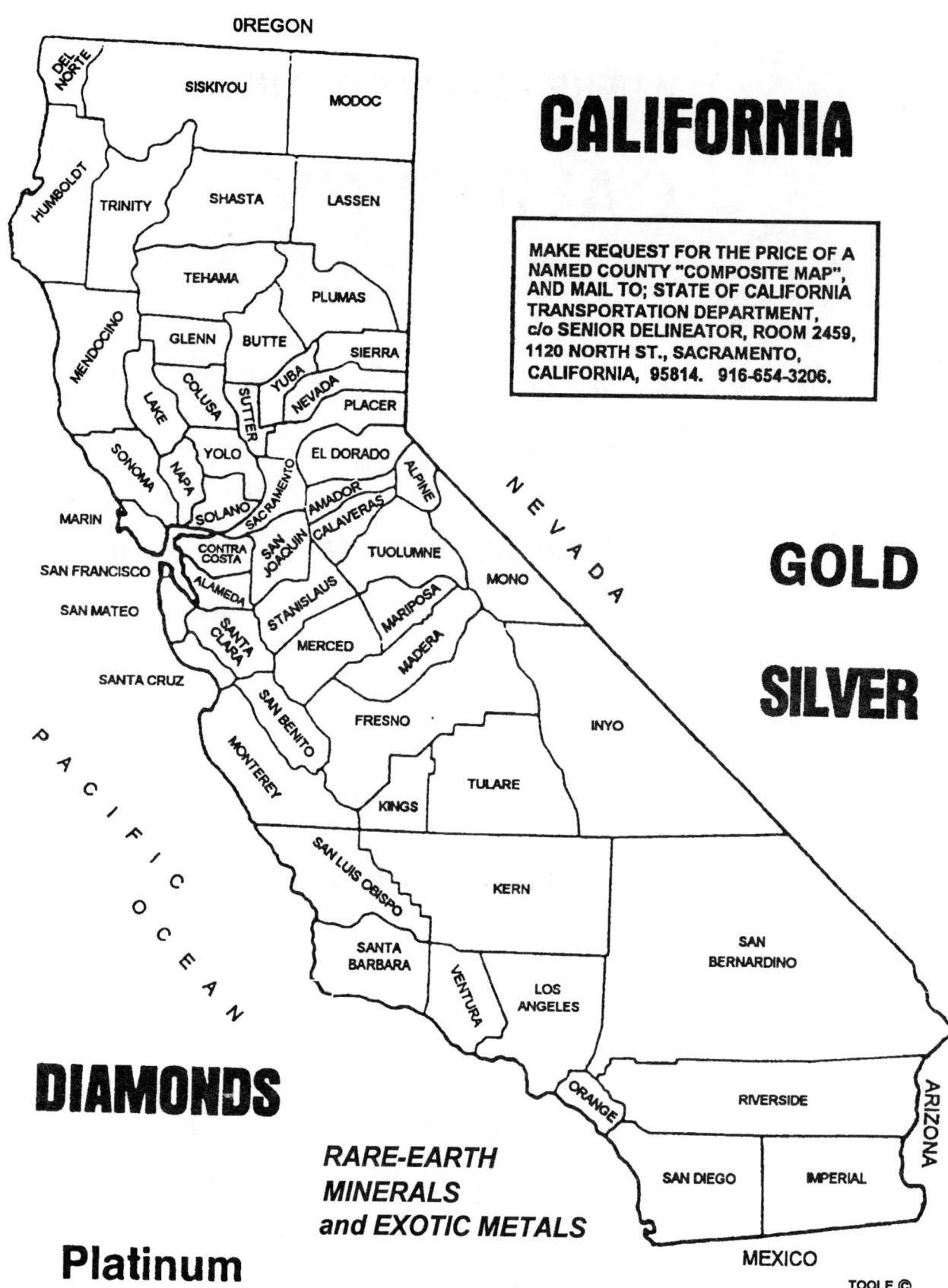

INDEX AND TABLE OF CONTENTS

TITLE   **PAGE NUMBERS**

ACKNOWLEDGMENT 1
AMADOR COUNTY With Placer Creeks.....................text 76.....map...... 76
AMERICAN MIDDLE FORK RIVER..............text 83.....maps 40 and...... 83
AMERICAN NORTH FORK RIVER.....text 81/82.....maps 40/81 and...... 83
AMERICAN SOUTH FORK RIVER..........text 79/80.....maps 79 and...... 80
BEAR RIVER...text 87.....map....... 87
BIG CHICO RIVER..text 100.....map..... 100
BIDWELL CREEK ....map.... 108
BIG FLAT Free-Use Site...text 61.....map...... 61
BUREAU OF LAND MANAGEMENT Addresses.......................text.......... 8
BOCA RESERVOIR..text 37.....map...... 38
BOGUS CREEK...text 91.....map...... 91
BOULDER CREEK Gold Site (Sacramento River)......text 73.....map...... 73
BULLARDS BAR Back-Country access.....................text 25.....map...... 26
BUTTE COUNTY Placer Rivers..................................text 99.....map....99
BUTTE CREEK Free-Use Site...............................text 101.....map....... 101
BUTTE COUNTY-Plumas County Placer Creeks...text 114.....map....... 115
CALIFORNIA DEPARTMENT OF FISH AND GAME Address.......text...... 9
CALAVERAS COUNTY Placer Creeks.....text 103-104.....maps..... 103-104
CARGO MOUNTAINS...text 133.....map... 133
CHEMICAL ANALYSIS TESTING KIT..text... 141
CHOCOLATE MOUNTAINS....................text 131-132.....maps... 131-132
CLASSIFICATION FOR DREDGING IN THE RIVERS.................text.. 9-18
CONVICT FLAT Free-Use Site.........................text 27/52.....maps.. 28/52
CONSUMNES RIVER................................(2) text 77.....(2) maps....... 77
DEER CREEK...text 109.....map...... 109
DEL NORTE COUNTY With Placer Creeks........text 64-66.....maps.. 64/67
DIAMONDS..text 105.....map....... 105
DISCLAIMER..text 2
DOG TOWN Free-Use Site....................................text 118.....map....... 118
DOUGLAS CITY...text.....59.....map...... 60
DREDGING IN THE RIVERS, CLASSIFICATION....................text........ 9-18
EEL RIVER...text 109.....map.......... 95
EL DORADO COUNTY With Placer Rivers............text 74.....map...... 74
FEATHER RIVER..........................text 102.....maps...... 102 / 111-112
FORESTHILL DIVIDE ROAD..................................text 40.....map.......... 40
FREE-USE GOLD MINING RECREATIONAL SITES
With Numbered Sites.....text 46-50.....maps.....49-55........ 61 / 63 / 118-119
FRENCH MEADOWS With Numbered Sites..........................text........... 42
FRESNO and MADERA COUNTY RIVERS..........text 117.....map........ 117

III

INDEX AND TABLE OF CONTENTS

TITLE   **PAGE NUMBERS**

GREENHORN RIVER..map....92
GROUSE RIDGE With Numbered Campsites..............text 21.....map....21
HUMBOLDT COUNTY With Gold Occurrences (Numbered Sites)...text.68
INDIAN VALLEY Free-Use Site.......................... text 27/51.....map...28/51
IMPERIAL COUNTY
With Gold Occurrences....................text 129-133.....maps.............130-133
JACKSON MEADOW With Numbered Campsites.........text 35.....map....36
JAMESTOWN (Gold Site Areas).....................text 97-98.....maps.....97-98
JULIAN-BANNER (Gold Placer Areas).....................text 134.....map....134
JUNCTION CITY (Gold Panning Sites)......................text.....57.....map..58
KERN COUNTY...text 119.....map....119
KEYSVILLE Gold Sites For Dry-Washing & Sluicing...text 119...map....119
KLAMATH RIVER With Gold Occurrences....text.57 / 69-71 / 89-91 / 94-95
                                                                                      maps ....57-89 / 91/ 94
LA PANZA RANGE With Rivers...............................text 124.....map....124
MADERA COUNTY..text 117.....map....117
MARIPOSA COUNTY With Gold-bearing Creeks.....text 113.....map.....113
MERCED COUNTY..text......120
MERCED RIVER..map.....120
MIDDLE FORK RIVER, AMERICAN........................text 83....map........83
MIDDLE NORTH YUBA RIVER.....................text 27/29.....maps........28/30
MIDWAY WELL (With Dry-Washing)..............text 132.....maps......131-132
MINES, OLD...text........37
MINING DISTRICTS, SOUTHERN CALIFORNIA..............text.......126/127
MODCO COUNTY..text..........108
MONTEREY COUNTY..text 116.....map....116
MOSQUITO RIDGE ROAD AREA With Back-Country Access.....text.....40
NATIONAL FOREST Addresses..text....6-7
NEVADA COUNTY AND RIVERS (With Gold Gravels)....text 86...map...86
NEW RIVER With Numbered Sites.............................text 72.....map....72
NORTH FORK MIDDLE FORK AMERICAN RIVER.....................map.....40
OREGON CREEK Free-Use Site........................ text 27/ 50.....map....28/50
OREGON, WHERE TO FIND GOLD...text....126
PIRU CREEK (Limited Access).....................text 121.....maps......121-122
PLACER COUNTY With Gold-Bearing Rivers..............text 84.....map.....84
PLATINUM (Six Member Family).......................text 106-107.....map.....107
PLUMAS COUNTY With Gold-Bearing Rivers..........text 111.....map.....115
POTHOLES (Dry-Washing and Gold Sniping).....text 129-130...map.....130
RAMSHORN Free-Use Site...............................text 27/ 53....map....28/53
RUBICON RIVER With Gold-Bearing Gravel...............text 78....map.......78

INDEX AND TABLE OF CONTENTS

TITLE	PAGE NUMBERS
SACRAMENTO RIVER	text 73.......map 74
SAN DIEGO COUNTY	text 134.....map 134
SAN GABRIEL EAST FORK RIVER	text 122.....map 123
SAN GABRIEL MOUNTAINS	text 122.....map 122
SANTA ANA MOUNTAINS AND CANYONS	text 125.....map 125
SANTA CLARA RIVER	text 121-122.....map 121-122
SANTA LUCIA RANGE	text 116.....map 116
SALINAS RIVER AND TRIBUTARIES	text 124.....map 124
SAUGUS-NEWHALL	text 122.....map.....122
SCOTT RIVER	text 90........map 90
SHASTA COUNTY AND RIVER	text 110......map 110
SHASTA RIVER (Limited Access)	text 93........map.......93
SIERRA BUTTES With Numbered Sites	text 31........map......32
SIERRAVILLE AND TRUCKEE With Numbered Sites	text 33/34...map 34
SOUTHERN CALIFORNIA MINING DISTRICTS	text....127-128
SOUTH YUBA RIVER With Numbered Sites	text 19/88.....maps ...20/88
SOUTH YUBA RIVER Free-Use Sites	text 62.....map 63
STAMPEDE With Numbered Sites	text 37......map......38
SUGAR PINE RESERVOIR Access To The Back-Country	text 44..map 45
TAHOE HIGHWAY # 89 Access To The Back-Country	map 39
TAHOE NATIONAL FOREST With Numbered Sites	text 23....map 24
TEXAS CANYON CREEK With Gold Particles	text 121.....maps 121-122
TRINITY COUNTY With Gold Occurrences	text 56....map 56
TRINITY RIVER With Gold-Bearing Gravel	text 60/109...maps.58/60/72
TRUCKEE CORRIDOR With Access To The Back-Country	text.......41
UNION FLAT RECREATIONAL FREE-USE SITE	text 54......map 54
VOLCANO CREEK PLACERS (Foresthill Area) (Limited Access)	text 84.....map 84
WARNER MOUNTAINS	text 108.....map.....108
WEBER CREEK Gold Placer Gravels	text 75.....map.......75
WHERE TO FIND GOLD IN OREGON With Maps 'n Directions	text 126
WILD PLUM RECREATIONAL FREE-USE SITE	text 29/55.....map 28/55
WILLOW CREEK With Access To The Back-Country	116.....map 116
WOODEN COVERED BRIDGE With Gold Site	map .28/50
WOODS CREEK PLACERS	text 97-98......maps 97-98
YUBA COUNTY	text 96........map..............96
YUBA RIVER	text 19 / 27 / 62 / 88.......maps.....20 / 28/ 30 / 63 / 88

v

ACKNOWLEDGMENT;

We gratefully acknowledge the many kind Persons who responded to our persistent pursuit in our quest for information from the many Federal, State, County agencies, including the generous assistance from Libraries, Universities and County Research Departments; and without their collaboration this work would not have been accomplished.

Extra thanks to Editors of friendly publications, writer friends and to concerned people who were always there with encouragement.

ALL RIGHTS RESERVED. No part of this publication may be stored, reproduced in a retrievable system or transmitted in any form or by means in electronics, mechanical, photo copying, recording or otherwise without prior written permission from the Author.

Library Of Congress catalog card number 94-061040. 94-061001.
Copyright 1993/1994 ISBN 0-9654559-0-4.
DELOS TOOLE, 5564 Lloyd CT SE, Salem. Oregon 97301

Infringements and plagiarism will be dealt aggressively with and severely under the protection of the National and International Copyright Laws.

DISCLAIMER;

The author, publisher and their associates take no responsibility for claims, lawsuits or litigation derived from the use of the furnished information presented within this Journal publication.

It is the responsibility of each user to be aware of lawful patented property, private property and legal mining claims before the reader enters into unfamiliar territory to search, prospect, dredge and to pan for valuable minerals.

In the pursuant of the supplied information the user will encounter mining locations with natural and artificial hazards that are detrimental and dangerous to limb. The user, by purchasing this publication further agrees to relieve the Author, Publisher and their associates from any liability incurred while engaged in pursuit of the information provided.

The Author and Publisher have no control over variables that are created by finicky bureaucratic regulatory influences, which may from time to time alter the authenticity of the supplied information found in this Journal, and therefore, does not guarantee an absolute.

TOPOGRAPHIC QUADRANGLES.
The following numbers correspond with the alphabetical Topographic Quadrangle names in the 7.5-minute maps. These numbers appear in a variety of symbols in approximate location on the hand drawn maps by the Author. Topographic map information may be obtained from; U.S. Geographical Survey, Box 25286, Federal Center, Bldg. 41, Denver, Colorado, 80225.

ACKERSON MOUNTAIN 1
ALDER PEAK 2
ALTURAS 3
AMERICAN HOME 4
ANGELS CAMP 5
ARAZ 6
AUBURN 7
AUKUM 8
BADGER MOUNTAIN 9
BALLOON DOME 10
BANGOR 11
BARD 12
BEAR RIVER RESERVOIR 13
BEN HUR 14
BIG ALKALI 15
BIG BAR 16
BLAKE MOUNTAIN 17
BOCA 18
BODIE 19
BOGUS MOUNTAIN 20
BRACEVILLE 21
BRANCH MOUNTAIN 22
BROWNS VALLEY 23
BUNKER HILL 24
BURNT PEAK 25
BURRO MOUNTAIN 26
CALAVERITAS 27
CAMINO 28
CAMPTONVILLE 29
CANBY 30
CAPE SAN MARTIN 31
CARBONDALE 32
CARRVILLE 33
CASCADE 34
CHALLENGE 35
CHEROKEE 36
CHERRY LAKE SOUTH 37
CHICAGO PARK 38
CHICKEN HAWK HILL 39
CHICO 40
CHIMNEY CANYON 41
CISCO GROVE 42
CLEMENTS 43
COHASSET 44
COLFAX 45
COLUMBIA 46
COPCO 47
CORONA SOUTH 48
COTTONWOOD 49
COVINGTON MILL 50
CRESSEY 51
CRESTON 52
CRYSTAL CRAG 53
CRYSTAL LAKE 54
DEL LOMA 55
DEVILS PEAK 56
DOGWOOD PEAK 57
DOG VALLEY 58
DOWNIEVILLE 59
DUCKWALL MOUNTAIN 60
DUNCAN PEAK 61
DUTCH FLAT 62
ELK GROVE 63
ENGLISH MOUNTAIN 64
ESCONDIDO 65
FELICIANA MOUNTAIN 66
FIDDLETOWN 67
FOLSOM 68
FORBESTOWN 69
FORESTHILL 70
FORT JONES 71
FORT BIDWELL 72
FRENCH CORRAL 73
FRENCH GULCH 74
CONTINUED ON NEXT PAGE

GALT	75
GASDEN	76
GARDEN VALLEY	77
GEORGETOWN	78
GENESEE VALLEY	79
GLAMIS	80
GLENDORA	81
GOLD LAKE	82
GRANITEVILLE	83
GRASS VALLEY	84
GREEK STORE	85
GREENVIEW	86
GROVELAND	87
HAMBURG	88
HAPPY CAMP	89
HASKINS VALLEY	90
HAYFORK	91
HAYPRESS VALLEY	92
HENNESSY PEAK	93
HETCH HETCHY RESERVR	94
HOBART MILLS	95
HOMEWOOD	96
HONCUT	97
HORNBROOK	98
HORSECAMP MOUNTAIN	99
HORSE CREEK	100
INDEPENDENCE	101
INDIAN CREEK BALDY	102
IRONSSIDE MOUNTAIN	103
JACKSON	104
JULIAN	105
JUNCTION CITY	106
KINSLEY	107
LAKE COMBIE	108
LATROBE	109
LA PANZA	110
LEWISTON	111
LOCKEFORD	112
LOMA RICA	113
MAD RIVER BUTTES	114
MARIPOSA	115
MARTIS PEAK	116
MERCED FALLS	117
MESA GRANDE	118
MICHIGAN BLUFF	119
MIDWAY WELL	120
MINT CANYON	121
MOCCASIN	122
MOKELUMNE HILL	123
MOKELUMNE PEAK	124
MONTAGUE	125
MOUNT SAN ANTONIA	126
MT. BIDWELL	127
MT. INGALLS	128
MT. SHASTA	129
NEVADA CITY	130
NEWHALL	131
NORDEN	132
NORTH BLOOMFIELD	133
OAKDALE	134
OREGON HOUSE	135
OROVILLE	136
OROVILLE DAM	137
PARADISE EAST	138
PARADISE WEST	139
PIKE	140
PILOT HILL	141
PINE GROVE	142
PIRU	143
PLACERVILLE	144
POLLOCK PINES	145
POZO SUMMIT	146
PROJECT CITY	147
PULGA	148
QUINCEY	149
RACKERBY	150
REDCREST	151
REDDING	152
RIVERTON	153
ROBBS PEAK	154
ROUGH & READY	155
ROYAL GORGE	156
SANTA YSABEL	157
SANTIAGO PEAK	158
SALYER	159
SARDINE PEAK	160

CONTINUED NEXT PAGE

SAWYERS BAR	161	TAHOE CITY	177
SCOTT BAR	162	TEN LAKES	178
SEIAD VALLEY	163	TIMBER KNOB	179
SHARKTOOTH PEAK	164	TUNNEL HILL	180
SHIPPEE	165	TUOLUMNE	181
SIERRA CITY	166	TURLOCK LAKE	182
SIERRAVILLE	167	TRUCKEE	183
SITTON PEAK	168	VALLEY SPRINGS	184
SLOUGHOUSE	169	WASHINGTON	185
SLY PARK	170	WEAVERVILLE	186
SMARTVILLE	171	WEBBER PEAK	187
SNELLING	172	WEED	188
SONORA	173	WEITCHPEC	189
STERLING	174	WESTVILLE	190
STORRIE	175	YREKA	191
STRAWBERRY VALLEY	176	YUBA CITY	192

THE MOTHER-LODE AREA.
The Author highly recommends for those, (and others), with limited vacation time to utilize the following suggestion; Golden Caribou Mining Club maintains several mining claims for its members to gold pan, dredge, sluice and high bank for placer gold on a variety of feeder creeks leading into the North Fk. Feather River and adjacent water-runs. Located at the confluence of the North Fork Feather River and the East Branch North Fork Feather River, (R.6E.-T.25N., sec. 19 & 20.) Located on St. hi-way no. 70 within Plumas County with location on the Plumas National Forest Map.

See pages 102 /112 / 142 of this book for additional information.

Write to; Golden Caribou Mining Club, P.O. Box 300, Belden, CA. 95915.   1-530-283-0956.

This area has been greatly influenced by the Rich Bar syndrome thrue the previous millennium of geological formation. The Feather River cuts thrue a 10 plus mile narrow canyon that intersects the host *serpentinite* and related ultramafic rocks that give indication of richly held gold which yearly enriches the adjacent water-runs.

NOTE: South of the E. Br. Fk. Feather River and So. of St hi-way no. 70 is ***off limits*** to mineral mining, (i.e.) "Bucks Lake Wilderness area."

There is indication of pre-ice Age conglomeration of alluvial gold hosted by high benches formed on the inside of the river bends. Most likely the aggregate concentrates are of the Eocene gold era where older and higher paystreakes exist. The area has been greatly influenced by the pre-ice Age laying down of alluvial gold, high or low aggregate strata, basins, ancient stream courses and pre-historic channels dominate the country-side that effect the adjacent water-runs with placer gold.

The Golden Caribou Mining Club members not only are allowed to keep all the gold they find, but, it affords each miner the ability to stay clear of private property and avoids offending claim owners in the area.

GOLD nugget sniping, panning for gold is a free-use activity within the National Forest Lands; there are exceptions in some areas where a small fee is charged and permits are required along with registration to control uses in sensitive areas. To obtain a Forestry Map and information for rules, regulations and registrations governing sluicing and suction dredging activity in a desired area of interest, write to;

ANGELES NATIONAL FOREST; 701 North Santa Antia Ave., Arcade, California, 921006.   818-574-1613.

CLEVELAND NATIONAL FOREST; 880 Front St., Room 5-N-4, San Diego, California, 92188.   619-557-5050.

ELDORADO NATIONAL FOREST; 100 Forni Road, Placerville, California, 95667.   916-644-6048.

INTO NATIONAL FOREST; 873 North Main St., Bishop, California, 93514.   619-873-5841.

KLAMATH NATIONAL FOREST; 1312 Fairlane Road, Yreka, California, 96130.   916-842-6131.

LASSEN NATIONAL FOREST; 55 So. Sacramento St., Susanville, California, 96130.   916-257-2151.

LOS PADRES NATIONAL FOREST; 6144 Calle Real, Galeta, California 93117.   805-683-6711.

MENDOCINO NATIONAL FOREST; 420 E. Laurel St., Willows, California, 95988.   916-934-3316.

MODOC NATIONAL FOREST; 441 North Main St., Alturas, California, 96101.   916-667-2247.

PLUMAS NATIONAL FOREST; P.O. Box 1500, Quincy, California, 95971.   916-283-2050.

SAN BERNARDINO NATIONAL FOREST; 1824 Commercenter, Circle, San Bernardino, California, 92408. 714-383-5588.

SEQUOIA NATIONAL FOREST; 900 W. Grand Ave., Porterville, California, 93257. 209-784-1500.

SHASTA-TRINITY NATIONAL FOREST; 2400 Washington Ave., Redding, California, 96001. 916-246-5222.

SIERRA NATIONAL FOREST; 1130 "O" St., Fresno, California, 93721. 209-487-5155.

SIX RIVERS NATIONAL FOREST; 507 "F" St., Eureka, California, 95501. 707-442-1721.

STANISLAUS NATIONAL FOREST; 19777 Greenley Road, Sonora, California, 95370. 209-532-3671.

TAHOE NATIONAL FOREST; Hwy. # 49 & Coyote St., Nevada City, California, 95959. 916-265-4531.

FOR AREAS OF INTEREST, THE FOLLOWING CALIFORNIA DEPTS. CAN BE OBTAINED BY WRITING TO;

To procure a California Highway Map, write to; State Of California Office Of Tourism, 1121 "L" St., Suite 103, Sacramento, Calif., 95814.

To obtain a County Map of a desired area, write to; Calif. Trans. Dept., P.O. Box 942874, Room 2459, Sacramento, CA., 94274. 916-654-3206.

For information about Mineral Withdrawal Areas in California, write to; California Dept. Of Parks & Recreation, P.O. Box 2390, Sacramento, Calif., 95811. 916-445-4624.

To obtain State Forestry Maps, write to; California Dept. Of Forestry, 1416 Ninth St., Sacramento, Calif., 95814. 916-445-9920.

For Mines & Mineral information write to; California Division Of Mines & Geology, P.O. Box 2980, Sacramento, Calif., 95812. 916-445-5716.

UNITED STATES DEPARTMENT OF INTERIOR BUREAU OF LAND MANAGEMENT STATE OFFICES;

There are numerous Resource Area Offices within each District Office. Contact the District Office for the area of interest, as the addresses change from-time-to-time with each of the Resource Area Office.

MAIN CALIFORNIA OFFICE;
CALIFORNIA STATE OFFICE;
Federal Building, Room E-2841, Cottage Way, Sacramento. California, 95825. 916-978-4754.

BAKERSFIELD DISTRICT OFFICE;
800 Truxtun Ave., Room 311, Bakersfield, Ca. 93301. 805-861-4191.
There are Four Resource Area Offices within this District:
1...BISHOP RESOURCE AREA.   2...CALIENTE RESOURCE AREA.
3...FOLSOM RESOURCE AREA.   4...HOLLISTER RESOURCE AREA.

SUSANVILLE DISTRICT OFFICE;
P.O. Box 1090, 705 Hall St., Susanville, CA., 96103.   916-257-5381.
There are Three Resource Area Offices within this District.
1...ALTURAS RESOURCE AREA   2...SURPRISE RESOURCE.
3...EAGLE LAKE RESOURCE AREA.

UKIAH DISTRICT OFFICE;
P.O. Box 1027, 555 Leslie St., Ukiah, CA., 95482. 707-462-3873.
There are Three Resource Offices within this District.
1...ARCATA RESOURCE AREA. 2...CLEARLAKE RESOURCE AREA.
3...REDDING RESOURCE AREA.

CALIFORNIA DESERT DISTRICT OFFICE;
6221 Box Springs Blvd., Riverside, CA., 92507. 714-697-5200.

Request BLM pamphlets; "Recreation Guide To BLM Public Lands". "Something For Everyone", (Western States). "California Camping Guide". "Campgrounds In California".

RESOURCES AGENCY OF CALIFORNIA DEPARTMENT OF FISH AND GAME SUCTION DREDGE REGULATIONS; (GENERAL).

NOTE; The overall structure of the General Regulations will remain the same with minor exceptions in changes from time-to-time. Best to check with the California Department Of Fish & Game. License & Revenue Branch, 3211 S St., Sacramento. CA., 95816. 916-653-4094.

EUREKA...619 2nd. St., Eureka, CA., 95501. 707-445-6493.
MENLO PARK 411 Burgess Dr., Menlo Park, CA., 94025. 415-688-6340.
MONTERY...2201 Garden Rd., Montery, CA., 93490. 408-649-2870.
SAN DIEGO...1350 Front St., Room #2041, San Diego, CA., 92101.
619-237-7311.

The use of a suction dredge within the water-runs of California requires a Dredge Permit for each dredge user and must be available upon demand by the area Official. Permits may be obtained from the following offices.

REGION 1...601 Locust St., Redding CA., 96001. 916-225-2300.
REGION 2...1701 Nimbus Rd., Rancho Cordova, CA., 95670.
916-355-0978.

REGION 3...7329 Silverado Trail, Napa, CA., 94558.
REGION 4...1234 E. Shaw Ave., Fresno, CA., 93710. 209-222-3761.
REGION 5...330 Golden Gate Shore #50, Long Beach, CA.,
213-590-5132.

Unless otherwise stated herein, the maximum dredge intake size permitted under a standard permit is 8 inches in diameter. Any dredge which has an intake nozzle or hose with an inside diameter that is larger than the permitted intake size, must have a constricting ring attached to the nozzle. This ring shall be of a solid one piece construction, with no openings other than the one intake opening of the allowable intake size. The constricting ring must be welded, or otherwise permanently attached, over the end of the intake nozzle; no quick release devices are permitted. The inner diameter of the intake hose may not be more than four inches larger than the permitted intake size.

DELOS TOOLE ©

LIST OF OPEN AND - OR CLOSED AREAS FOR USE WITH A STANDARD PERMIT, (Special Permits Not Valid in These Waters Unless So Specified In The Special Permit). A Special Permit must be obtained from the Department in order to operate a suction dredge in non-conformance with these regulations.

Suction Dredging into the bank of any stream is prohibited.

CALIFORNIA is divided into eight classes which are listed below with the dates when suction - vacuum dredging is permitted in each class.

*CLASS...A...*Closed Waters...No dredging permitted at any time.
*CLASS...B...*Open to dredging from July 1 through August 31.
*CLASS...C...*Open to dredging from the fourth Saturday in May through October 15.
*CLASS...D...*Open to dredging from from July 1 through September 15.
*CLASS...E...*Open to dredging from July 1 through September 30.
*CLASS...F...*Open to dredging from December 1 through June 30.
*CLASS...G...*Open to dredging from the fourth Saturday in May through September 30.
*CLASS...H...*Open to dredging throughout the year.

County areas are included in one or more of the above classes. In Addition to the classes listed opposite Counties below, most Counties have some further detailed restrictions or additional open-waters. These further restrictions or additional open-waters are listed alphabetically on the following pages by stream or water-runs with the particular applicable County shown by parentheses. (Check this list of water-runs before dredging).

ALAMEDA...Class H.
ALPINE...All waters Class C.
AMADOR...East of Hwy. # 49 is Class C, remainder of area is Class H.
BUTTE...Class C.
CALAVERAS...East of Hwy. # 49 is Class C, remainder of area is Class H.
COLUSA...Class H.
CONTRA COSTA...Class H.
DEL NORTE ...Class E.
EL DORADO...East of Hwy. # 49 is Class C, remainder of area is Class H.
FRESNO...Within the external boundaries of the National Forest is Class C, the remainder of the area is Class H.
GLENN...Class H.
HUMBOLDT...Class E.
IMPERIAL...Class H.
INYO...Class A.

KERN...The Kern River & tributaries from Isabella Dam upstream is Class A, the remainder of the area is Class H.
KINGS...Class H.
LAKE...Class H.
LASSEN...Class C.
LOS ANGELES...Class H. Except where noted.
MADERA...Within the external boundaries of the National Forests is Class C, the remainder of the area is Class H.
MARIN...Class A.
MARIPOSA...Within the external boundaries of the National Forest is Class C, the remainder of the area is Class H.
MEDOCINO...Class A.
MERCED...Class H.
MODOC...Class C.
MONO...Class A.
MONTEREY...Class A
NAPA...Class A.
NEVADA...East of Hwy. # 49 is Class C, the remainder of the area is Class H.
ORANGE...Class H.
PLACER...East of Hwy. # 49 is Class C, the remainder of the area is Class H.
PLUMAS...Class C.
RIVERSIDE...Class H.
SACRAMENTO...Class H.
SAN BENITO...Class A.
SAN BERNARDINO...Class H. Except where noted.
SAN DIEGO...Class H.
SAN FRANCISCO...Class H. Except where noted.
SAN JOAQUIN... Class H.
SAN LUIS OBISPO...Class A.
SAN MATEO...Class A.
SANTA BARBARA...Class H.
SANTA CLARA...Class H.
SANTA CRUZ...Class A.
SHASTA...Class C.
SIERRA...Class C.
SISKIYOU...Class E.
SOLANO...Class H.
SONOMA...Class A.
STANISLAUS...Class H.
SUTTER...Class H.
TEHAMA...Class D.

TRINITY...Class E.
TULARE...Within the external boundaries of the National Forest is Class C, the remainder is Class H.
TUOLUMNE...East of Highway # 49 is Class C, the remainder is Class H.
VENTURA...Class H.
YOLO...Class H.
YUBA...Class H.

CLASS REGULATIONS WITH ADDITIONAL WATERS AND AREAS;

AMERICAN RIVER (SACRAMENTO COUNTY). The main stem American River from the Sacramento River upstream to Nimbus Dam is Class A.
AMERICAN RIVER, MIDDLE FORK (EL DORADO AND PLACER COUNTIES). The main stem American River Middle Fork from its junction with the North Fork of the American River upstream to the confluence with the Rubicon River is Class C. Recreational dredging is allowed in the Auburn State Recreational Area on an interim management basis. Contact the Auburn State Recreational Area for info.
AMERICAN RIVER, NORTH FORK (PLACER COUNTY). The main stem North Fork American River from Folsom Reservoir to the Colfax-Iowa Hill Road Bridge is Class C. From the Colfax-Iowa Hill Road Bridge upstream to Heath Springs at location T16N R14E Sec. 26, is Class A.
AMERICAN RIVER, SOUTH FORK (EL DORADO COUNTY). The main stem South Fork American River from Folsom Reservoir upstream to the Highway No. 49 bridge at Coloma is Class C.
AMERICAN RIVER, SOUTH FORK TRIBUTARIES (EL DORADO COUNTY). All tributaries to the South Fork American River from Folsom Reservoir upstream is Class C.
ANTELOPE CREEK AND TRIBUTARIES (PLACER COUNTY). Antelope Creek and its tributaries are Class B.
AUBURN RAVINE AND TRIBUTARIES (PLACER COUNTY). Auburn Ravine and its tributaries are Class B.
BEAR RIVER (PLACER COUNTY). The main stem Bear River from Forty Mile Road to the South Sutter Irrigation District's diversion Dam is Class D.
BIG CHICO CREEK (BUTTE COUNTY). The main stem Big Chico Creek from Manzanita Avenue in Chico to the head of Higgens Hole at location T24N R3E Sec. 31 is Class A.
BIG CREEK AND TRIBUTARIES (FRESNO COUNTY). Big Creek, tributary to the Kings River, and its tributaries are Class A.
BIG CREEK (TRINITY COUNTY). The main stem Big Creek is Class A.
BUTTE CREEK (BUTTE COUNTY). The main stem Butte Creek from the Sutter County Line upstream to the Durham-Oroville Highway Bridge is Class H, from the Durham-Oroville Highway Bridge upstream to the intake of Centerville Ditch at location T23N R3E Sec 10 is Class A.

BLUE CREEK AND TRIBUTARIES (DEL NORTE AND HUMBOLDT COUNTIES). Blue Creek and its tributaries are Class A.

CALAVERAS RIVER AND TRIBUTARIES (CALAVERAS AND SAN JOAQUIN COUNTIES). The Calaveras River and its tributaries below New Hogan Reservoir are Class B.

CANYON CREEK (YUBA COUNTY). The main stem Canyon Creek from its mouth upstream to Sierra-Yuba County line, (T20N R8E Sec. 25), is Class C.

CHERRY CREEK (TUOLUMNE COUNTY). The main stem of Cherry Creek is Class B.

CHOWCHILLA RIVER (MADERA AND MARIPOSA COUNTIES). The main stem Chowchilla River from Eastman Lake upstream to the West and East forks of the Chowchilla River is Class A.

CHOWCHILLA RIVER WEST FORK (MADERA AND MARIPOSA COUNTIES). The main stem West Fork Chowchilla River from its mouth upstream to Highway no. 49 bridge is Class A.

CLAVEY RIVER (TUOLUMNE COUNTY). The main stem Clavey River is Class A.

CLEAR CREEK AND TRIBUTARIES (SISKIYOU COUNTY). Clear Creek and Tributaries are Class A.

COLORADO RIVER AND TRIBUTARIES (IMPERIAL, RIVERSIDE AND SAN BERNARDINO COUNTIES). The main channel and all sloughs and tributaries of the Colorado River are Class A.

COSUMNES RIVER (SACRAMENTO, AMADOR AND EL DORADO COUNTIES). The main stem Cosumnes River from the Western Pacific Railroad Bridge to about 1/4-mile above the mouth upstream to the Latrobe Highway Bridge is Class D. From the Latrobe Highway Bridge upstream to the confluence with the North and Middle Forks of the Cosumnes River is Class H.

COSUMNES RIVER, NORTH FORK (EL DORADO COUNTY). The main stem North Fork Cosumnes River from the Middle Fork of the Cosumnes River upstream to Somerset-Pleasant Valley Road Bridge is Class H.

COSUMNES RIVER, MIDDLE FORK (EL DORADO COUNTY). The main stem Middle Fork Cosumnes River from North Fork Cosumnes River upstream to Bakers Ford on the Aukum-Somerset Road is Class H.

COSUMNES RIVER, SOUTH FORK (AMADOR AND EL DORADO COUNTIES). The main stem South Fork Cosumnes River from Middle Fork Cosumnes River upstream to the County Road Bridge at River Pines is Class H.

COW CREEK AND TRIBUTARIES (FRESNO COUNTY), are Class A.

CURTIS CREEK (TUOLUMNE COUNTY), is Class C.

DEEP CREEK (SAN BERNARDINO COUNTY), is Class A.

DEER CREEK (NEVADA COUNTY). The main stem Deer Creek from Ponderosa Way below Rough and Ready Falls (T16N R7E Sec 13), upstream to Highway no. 49 is Class C.

DILLION CREEK AND TRIBUTARIES (SISKIYOU COUNTY), are Class A.

DINKEY CREEK AND TRIBUTARIES (FRESNO COUNTY), are Class A.

EAGLE CREEK (TUOLUMNE COUNTY), is Class C.
EASTMAN LAKE (MADERA AND MARIPOSA COUNTIES), is Class A.
EEL RIVER, ALL FORKS AND TRIBUTARIES (MENDOCINO COUNTY).
The Eel River, all forks and its tributaries upstream of the Humboldt-Mendocino and Trinity-Mendocino County lines are Class A.
EEL RIVER, MIDDLE FORK AND TRIBUTARIES (MENDOCINO AND TRINITY COUNTIES), are Class A.
FEATHER RIVER (BUTTE COUNTY). The main stem Feather River from Honcut Creek (T17N R3E Sec. 27) upstream to the Highway # 70 Bridge is Class B, and from the Highway # 70 Bridge upstream to Oroville Dam is Class A.
FEATHER RIVER, SOUTH FORK (BUTTE AND PLUMAS COUNTIES).
The main stem South Fork Feather River from Oroville Reservoir upstream to Little Grass Valley Dam (T22N R9E Sec. 31) is Class C.
FLAT CREEK AND TRIBUTARIES (SHASTA COUNTY), are Class H.
FRENCH CREEK (TRINITY COUNTY), is Class A.
GRAPEVINE CREEK (TUOLUMNE COUNTY, is Class B.
HORTON CREEK (TUOLUMNE COUNTY), is Class A.
HUNTER CREEK (TUOLUMNE COUNTY), is Class B.
INDEPENDENCE CREEK AND TRIBUTARIES (NEVADA AND SIERRA COUNTIES), from Independence Lake upstream is Class A.
JAWBONE CREEK (TUOLUMNE COUNTY), is Class B.
KAWEAH RIVER (TULARE COUNTY), upstream of Kaweah Reservoir is Class A.
KERN RIVER AND TRIBUTARIES (KERN AND TULARE COUNTIES), from the Isabella Dam upstream are Class A.
KERN RIVER, SOUTH FORK AND TRIBUTARIES (KERN AND TULARE COUNTIES), are Class A.
KING RIVER AND TRIBUTARIES (FRESNO AND KINGS COUNTIES), from Tulare Lake upstream to Pine Flat Dam is Class A.
KLAMATH RIVER, MAIN STEM (DEL NORTE, HUMBOLDT AND SISKIYOU COUNTIES). From the mouth upstream to the Salmon River is Class G, from the Salmon River upstream to 500 feet downstream of the Scott River is Class H, from 500 feet downstream of the Scott River upstream to Iron Gate Dam is Class G, and from Iron Gate Dam to the Oregon Border is Class A.
KNIGHTS CREEK (TUOLUMNE COUNTY), is Class C.
LAVEZZOLA CREEK (SIERRA COUNTY), is Class C.
LITTLE ROCK CREEK AND TRIBUTARIES (LOS ANGELES COUNTY), from the Sycamore Campground in the Los Angeles National Forest upstream are Class A.
LITTLE SWEDE CREEK (TRINITY COUNTY), is Class A.
MACKLIN CREEK (NEVADA COUNTY), from its confluence with the Middle Fork Yuba River (T19N R12E Sec. 16 upstream is Class A.
MALIBU CREEK AND TRIBUTARIES (LOS ANGELES COUNTY), is Class A.
McCLOUD RIVER (SHASTA COUNTY), from the southern boundary of T38N R3W Sec. 16, upstream to Lake McCloud Dam is Class A.

MERCED RIVER (MERCED COUNTY). From the San Joaquin River upstream to Crocker-Huffman Dam, upstream from Snelling, is Class A.
MERCED RIVER (MARIPOSA COUNTY), is Class C.
MERCED RIVER, NORTH FORK (MARIPOSA COUNTY), is Class C.
MINER'S RAVINE AND TRIBUTARIES (PLACER COUNTY), are Class B.
MINNOW CREEK (TUOLUMNE COUNTY), is Class A.
MOKELUMNE RIVER (AMADOR, CALAVERAS AND SAN JOAQUIN COUNTIES), from Burella Road upstream to Camanche Dam is Class A, from Camanche Dam upstream to Pardee Dam is Class H, and from Pardee Dam upstream is Class C.
MUD CREEK (BUTTE COUNTY), from Big Chico Creek upstream is Class C.
NELSON CREEK (PLUMAS COUNTY), is Class C.
NEW RIVER AND TRIBUTARIES (TRINITY COUNTY), upstream from the East Fork New River are Class A.
NEW RIVER EAST FORK AND TRIBUTARIES (TRINITY COUNTY), from the New River upstream are Class A.
PIRU CREEK AND TRIBUTARIES (VENTURA AND LOS ANGELES COUNTIES), are Class A.
PIT RIVER AND TRIBUTARIES (LASSEN AND MODOC COUNTIES), Class A.
POOR MAN CREEK AND TRIBUTARIES (TUOLUMNE COUNTY), are Class A.
PORTUGUESE CREEK AND TRIBUTARIES (MADERA COUNTY), are Class A.
ROCK CREEK (BUTTE COUNTY), from Big Chico Creek upstream to the Butte, Tehama County line is Class C.
ROCK CREEK AND TRIBUTARIES (SHASTA COUNTY), are Class H.
ROSE CREEK (TUOLUMNE COUNTY), is Class C.
RUBICON RIVER AND TRIBUTARIES (EL DORADO AND PLACER COUNTIES). The Rubicon River and its tributaries are Class C. Dredges with an intake larger than four-inches is prohibited.
SACRAMENTO RIVER AND TRIBUTARIES (SEVERAL COUNTIES) from the San Francisco Bay upstream to Shasta Dam is Class A. From Shasta Lake upstream to Box Canyon is Class A.
SALMON RIVER (SISKIYOU COUNTY), is Class D.
SALMON RIVER, NORTH FORK (SISKIYOU COUNTY), from the South Fork Salmon River upstream to the Marble Mountain Wilderness boundary is Class D.
SALMON RIVER, SOUTH FORK (SISKIYOU COUNTY), from the North Fork Salmon River upstream to the Trinity Alps Wilderness boundary is Class D.
SALT CREEK AND ITS TRIBUTARIES (RIVERSIDE COUNTY) are Class A.
SAN FELIPE CREEK AND TRIBUTARIES (IMPERIAL AND SAN DIEGO COUNTIES), are Class A.
SAN GABRIEL RIVER, EAST FORK AND TRIBUTARIES (LOS ANGELES COUNTY), from Cattle Canyon upstream is Class A.
SAN GABRIEL RIVER, WEST FORK AND TRIBUTARIES (LOS ANGELES COUNTY), from the Rincon Guard Station upstream is Class A.

SAN JOAQUIN RIVER (SEVERAL COUNTIES), from the Delta upstream to Friant Dam (Millerton Lake), is Class A.
SAN JUAN CREEK AND TRIBUTARIES (ORANGE AND RIVERSIDE COUNTIES), from its mouth upstream is Class A.
SAN MATEO CREEK AND TRIBUTARIES (SAN DIEGO, ORANGE AND RIVERSIDE COUNTIES), from its mouth upstream is Class A.
SANTA ANA RIVER AND ITS TRIBUTARIES (SAN BERNARDINO COUNTY), from the mouth of Bear Creek upstream is Class A.
SANTA CLARA RIVER AND TRIBUTARIES (LOS ANGELES AND VENTURA COUNTIES), from the Los Angeles-Ventura County line upstream are Class A, except that Texas Canyon Creek is Class H.
SANTIAGO CREEK AND TRIBUTARIES (ORANGE COUNTY) within the Cleveland National Forest is Class A.
SAXON CREEK (MARIPOSA COUNTY), is Class A.
SCOTT RIVER AND TRIBUTARIES (SISKIYOU COUNTY), is Class G.
SECRET RAVINE AND TRIBUTARIES (PLACER COUNTY), are Class B.
SESPE CREEK (VENTURA COUNTY), from the Los Padres National Forest boundary upstream to its confluence with Tule Creek is Class A.
SHAY CREEK & TRIBUTARIES (SAN BERNARDINO COUNTY), is Class A.
SHASTA RIVER AND TRIBUTARIES (SISKIYOU COUNTY), are Class A.
SHERLOCK CREEK (MARIPOSA COUNTY), is Class A.
SILVER KING CREEK AND TRIBUTARIES (ALPINE COUNTY), are Class A.
SIX-BIT CREEK AND TRIBUTARIES (TUOLUMNE COUNTY), are Class A.
SMITH RIVER MIDDLE FORK (DEL NORTE COUNTY), is Class D.
STANISLAUS RIVER (CALAVERAS, SAN JOAQUIN, STANISLAUS AND TUOLUMNE COUNTIES). From the San Joaquin River upstream to Goodwin Dam is Class A. From New Melones Dam upstream, excluding Melones Reservoir is Class C.
SULLIVAN CREEK (TUOLUMNE COUNTY), is Class C.
SUTTER CREEK (AMADOR COUNTY), from Highway # 49 upstream to Pine Gulch Road is Class H.
SYCAMORE CREEK AND TRIBUTARIES (FRESNO COUNTY), are Class A.
TEXAS CANYON CREEK (LOS ANGELES COUNTY), is Class H.
TRINITY RIVER, MAIN STEM BELOW LEWISTON DAM (HUMBOLDT AND TRINITY COUNTIES), from the Klamath River upstream to the South Fork Trinity River is Class A. From the South Fork Trinity River upstream to the North Fork Trinity River is Class H. From the North Fork Trinity River upstream to Grass Valley Creek is Class D. From Grass Valley Creek upstream to Lewiston Dam is Class A.
TRINITY RIVER, MAIN STEM AND TRIBUTARIES ABOVE LEWISTON DAM (TRINITY COUNTY), are open to dredging from July 1 thru October 15.
TRINITY RIVER, NORTH FORK AND TRIBUTARIES (TRINITY COUNTY), upstream from Hobo Gulch Campground is Class A.

TUOLUMNE RIVER (STANISLAUS COUNTY), from the Waterford Bridge upstream to La Grange Dam is Class A.
TUOLUMNE RIVER, NORTH FORK, AND TRIBUTARIES (TUOLUMNE COUNTY), are Class B.
TURNBACK CREEK AND TRIBUTARIES (TUOLUMNE COUNTY), are Class A.
WOLF CREEK (NEVADA COUNTY), from the Tarr Ditch Diversion (T15N R8E Sec. 10) upstream is Class C.
WOODS CREEK AND TRIBUTARIES (TUOLUMNE COUNTY), from Harvard Mine Road (Jamestown) downstream are Class C. From Harvard Mine Road upstream is Class A.
WOOLEY CREEK AND TRIBUTARIES (SISKIYOU COUNTY), are Class A.
YUBA RIVER (YUBA COUNTY), from its mouth at Marysville upstream to Highway no. 20 is Class B. From Highway no. 20 upstream to Englebright Dam is Class A.
YUBA RIVER, NORTH FORK (SIERRA AND YUBA COUNTIES), from the Middle Fork of the Yuba River upstream to Fiddle Creek is Class H.

Rules and regulations change frequently by authorities due to environmental conditions changing from stable to sensitive areas by over-usage. It is the responsibility of the reader to pursue the up-grades for dredging rules of those sensitive water-runs. Contact your local office of the Department of Fish and Game for detailed information.

Wild and Scenic Rivers include portions of the American River (North Fork and Lower American River), Big Sur River, Eel River, Feather River, Kern River, Kings River, Klamath River, Merced River, Sespe River, Sisquoc River, Smith River, Trinity River and the Tuolumne River. Dredging may be restricted on Federally designated Wild and Scenic Rivers. Contact the Federal Land Managing Agency for details.

Waters in National Parks, National Monuments, State Parks and designated Wilderness Areas are closed by the land managing agencies.

Some waters in the San Gabriel Mountains are closed. Contact the Angeles National Forest Office before dredging.

Portions of the Sequoia and Sierra National Forests, designated as the Kings River Special Management Area, are closed to dredging. Contact the appropriate U.S. Forest Service Office for details.

The Auburn State Recreation Area has special restrictions on suction dredging. Contact the Auburn State Recreation Area Office for details. There are open areas within the Auburn State Recreation Park Area that are set aside for Recreation Gold Mining as Free-use Sites. Check this out as this is a lucrative gold-bearing area.

SPECIAL CONDITIONS RELATING TO WILD & SCENIC RIVERS.

Certain rivers in California have been designated as components of the Federal Wild & Scenic River Systems or as study rivers for possible inclusion in the Federal System. Restrictions have been placed on **mining** in Federal Lands or within 1/4 mile of water-runs within this classification.

KERN RIVER, (Tulare County). 5,600 feet upstream from the Johnsondale Bridge to the headwaters.

MERCED RIVER. From the upper end of Lake McClure at the 867 foot waterlevel, the high water level, upstream to the River's headwaters.

MIDDLE FORK FEATHER RIVER. From Lake Oroville to the town of Backworth. **Mining is prohibited in the wild section of the river.** The USFS has other specific regulations regarding mining claims within this section. Contact Plumas National Forest, P.O. Bx 1500, Quincy, California, 95971.

NORTH FORK AMERICAN RIVER. From the upstream limits of Auburn Reservoir to the Cedars. **ALL NEW MINING IS PROHIBITED.**

TUOLUMNE RIVER. From Don Pedro Reservoir upstream to the Yosemite Park Boundary.

DELOS TOOLE ©

TAHOE NATIONAL FOREST-SOUTH YUBA RIVER- CALIF. HWY. #20

Highway #20 - South Yuba River

Highway #20 runs eastward from Nevada City 26 miles to its intersection with Interstate 80. It provides access to the South Yuba River in the vicinity of the town of Washington. Activities include camping, picnicking, hunting and viewing scenery, with the additional pleasures of fishing, swimming, and **gold panning in the South Yuba River area**. Supplies are available in Nevada City and Washington. Normal use season is May through October.

1...White Cloud Campground. Located 12 miles east of Nevada City on Hwy. 20. 46 tent and trailer campsites. Flush & vault toilets. Piped water. Good parking. Easy Hwy. access. Close to Nevada City. Adjacent to White Cloud Fire Station.

2...White Cloud Picnic Area. Day use only. Located on Hwy. 20 across from White Cloud Campground. 17 picnic areas with fireplaces & fire rings. Flush toilets. Piped water. Good parking. Facilities for both family and group picnics. Easy Highway access. Close to Nevada City.

3...Skillman Campground. Located on Hwy. 20, 15 miles east of Nevada City. 16 campsites with 12 spaces for trailers. Vault toilets. Piped water. Limited parking. Easy Highway access.

4...Keleher Picnic Area. Day use only. Located on the South Fork of the Yuba River, 2 miles upstream from the town of Washington. 9 picnic sites with tables and stoves. Vault toilets. Piped water. Plenty of parking. Adjacent to the Yuba River. **Gold panning. Nugget sniping with a metal detector.**

5...Golden Quartz Picnic Area. Day use only. Located on the South Fork of the Yuba River, 3 miles upstream from Keleher Picnic Area on a rough winding road. 7 picnic sites with tables and grills. Vault toilet. Adequate parking. No garbage receptacle (pack in pack out refuse). Stream water source (purify before drinking). Adjacent to Yuba River. Remote location. **Gold panning & nugget sniping.**

6...Rock Creek Nature Study Area. Day use only. Located 4.5 miles east of Nevada City on Hwy. 20, and 1.5 miles northeast on Rock Creek Road. Rough & winding road. (Use this study trail to explore adjacent water-runs in the back-country for gold panning and nugget sniping with metal detector). One-mile long self-guided nature study trail. At the head of trail, pick up a brochure that explains the natural features of the Rock creek area. 3 picnic tables at trailhead. Vault toilet. Ample parking. No piped water. Purify all drinking water.

GOLD PANNING AND NUGGET SNIPING ON THE SOUTH FORK OF THE YUBA RIVER. Easy access to back-country water-runs with Highway # 20.

Tahoe National Forest, Hwy. # 49 and Coyote St., Nevada City, CA. 94111. 916-265-4531. Check for additional information and pick up Forestry Map.

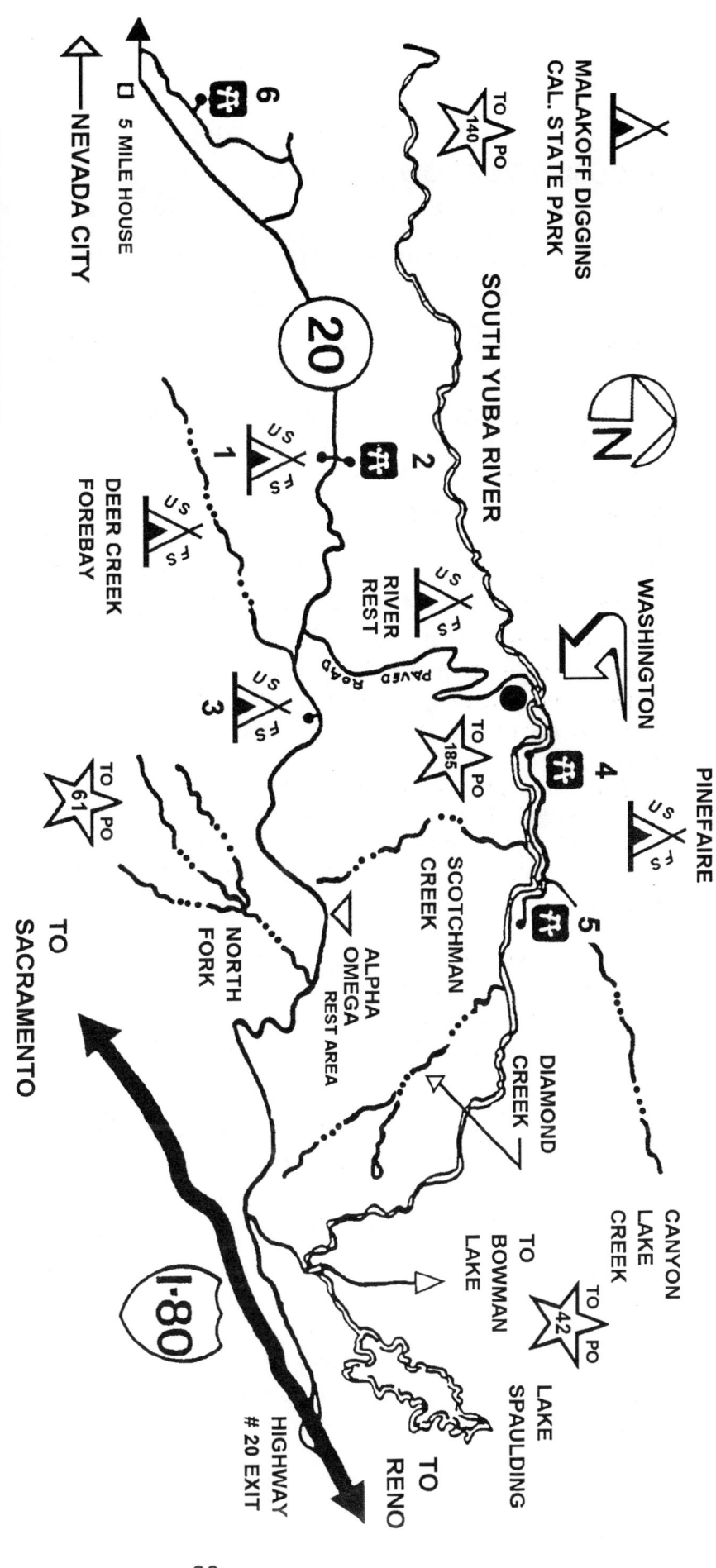

TAHOE NATIONAL FOREST ROAD AREA
GROUSE RIDGE-BOWMAN LAKE

Bowman Road Area (Grouse Ridge-Bowman Lake)

The Bowman Road proceeds north from Hwy. # 20 for 3.5 miles west of the I-80 intersection. It travels north 16 miles to Bowman Lake and then northeast 6 miles to the Jackson Meadow area. Activities include stream and lake fishing, camping, hiking, swimming, picnicking, boating and hunting in season. Supplies available in Nevada City or along Interstate 80. Normal season use is June through Oct.

7...Fuller Lake Campground. Located on Fuller Lake, 4 miles north of Hwy. # 20. Available 9 camp sites. Vault toilets. Limited parking for trailers & cars outside camping units. No piped water. Lake water available (purify before using).

8...Grouse Ridge Campground. Located near Grouse Ridge lookout, 11 miles north of Hwy. # 20. Campsites (9) with 5 spaces for small trailers. Vault toilets. Piped water. Plenty of parking at campground & trailhead. Pack-in/pack-out. No garbage receptacle. Access to the back-country for **gold panning** and test hole exploration of water-runs.

It is recommended that for access to the following back-country campgrounds, pssenger vehicles and trailers use the route from Graniteville. Take Hwy. # 49 to Tyler Foote Crossing, (look for mine tailings along this route), and proceed east through North Columbia, approximately 28 miles from Hwy. # 49 to Bowman Lake.......

9...Bowman Lake Campground. Located at Bowman Lake, 16 miles north of Hwy. # 20. Undesignated (7) sites. Vault toilets. Parking available. Pack-in/pack out (no garbage cans). Lake water source (purify before using). Fishing, swimming, boating and adjacent to back-country water-runs for **gold panning, nugget sniping,** test hole exploration at remote sites.

10...Jackson Creek Campground. Located on Jackson Creek, 18 miles north of Hwy. # 20. Campsites (14). Vault toilets. Limited parking. Campsites (14). Vault toilets. Limited parking. Pack-in/pack-out (no garbage cans). Purify water before using. Remote location, adjacent to back-country water-runs for **gold panning.**

11...Canyon Creek Campground. Located on Canyon Creek below Faucherie Reservoir, 3 miles southeast of Jackson Creek campground on Forest Road no. 18N13. Prairie Creek Crossing can be difficult at times for passenger vehicles, recreational vehicles and trailers. Campsites (20), space for 11 trailers. **Remote campsite.** Limited parking. Vault toilets. Piped in water. Pack-in/pack-out.

12...Faucherie Group Campground. Located at Faucherie Reservoir, 1 mile above Canyon Creek Campground on Forest Road no. 18N13. Reservations required through the Nevada City Ranger District, 12012 Sutton Way, Grass Valley, CA., 95945. 916-273-1371. Group campsites with tables & stoves, (25 people) . Remote location adjacent to back-country water-runs for **gold panning** and exploration testing. Vault toilets. Limited parking. Limited space for small trailers.

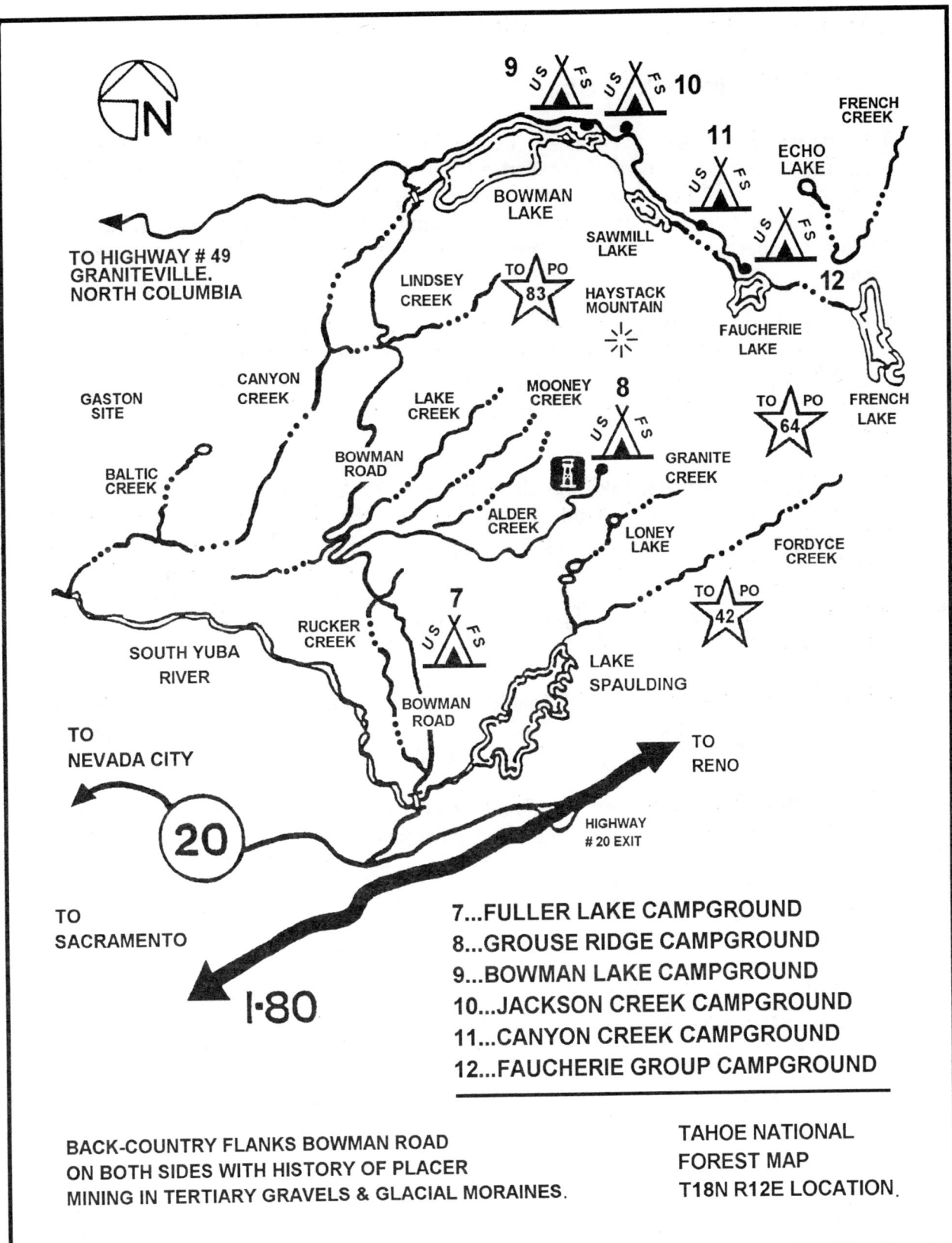

TAHOE NATIONAL FOREST - INTERSTATE HIGHWAY # 80

INTERSTATE HIGHWAY # 80;

Interstate # 80 crosses the western Forest boundary near Blue Canyon and goes eastward up the South Yuba River Canyon to Donner Summit. It continues to the east through Truckee and down the Truckee River Canyon into Nevada. The following sites are all adjacent to or accessible from Interstate # 80. Amenities are available at Emigrant Gap, Cisco Grove, Soda Springs, Norden and Truckee. Use season June thru October.

13...North Fork Campground. Take the Emigrant Gap exit off I- # 80 and travel 6 miles southeast on the Texas Hill Road. Remote location with piped water. Vault toilet. Limited space for small trailers. Limited parking. Adjacent to North Fork of the North Fork of the American River. Back-country offers **gold panning** and exploration with tests in gravel beds.

14...Tunnel Mills Group. Take the Emigrant Gap exit off I- #80 and travel 7.5 miles southeast on the Texas-Hill Road. This group campground will require reservation for groups. Contact; Nevada City Ranger District, 12012 Sutton Way, Grass Valley, CA., 95945. 916-273-1371. Two campsites for large groups with tables and grills. Vault toilets. No piped water. Limited parking. Stream water source, purify before drinking. Adjacent to East Fork of the North Fork of the American River. **Remote location.** Access to the back-country water-runs for **gold panning.**

15...Indian Springs Campground. Take the Eagle Lakes exit off I #80 and turn north for 0.5 miles. Space for trailers with 35 campsites. Piped water. Vault toilets. Adjacent to South Fork Yuba River. Good jumping off point into the back-country.

16...Woodchuck Campground. Take the Cisco Grove exit off I #80 and turn north. Turn left onto frontage road after crossing over the freeway and then turn right just before Thousand Trails onto Rattlesnake Road. Proceed 3 miles to campground sites. Limited parking with 8 tent campsites. Vault toilets. Stream water source, purify before drinking. Pack-in/pack-out, no garbage cans. Adjacent to Rattlesnake Creek and back-country water-runs for **gold panning.**

17...Sterling Lake Campground. Using the directions to Woodchuck Campground above, proceed 3.5 miles past Woodchuck Campground to Sterling Lake Campground and to the back-country water-runs for gold panning, metal detecting and gold nugget sniping. Steep-winding road. Not suited to trailers. Remote location. Vault toilets. 6 campsites. Pack-in/-pack-out, no garbage cans. Lake water source, purify before drinking. Adjacent **back-country** water-runs.

18...Big Bend Campground. From eastbound I #80, take the Big Bend exit and continue east 1/4 mile to campsite; from westbound I #80, take the Rainbow exit and travel west for 1.5 miles to campground. 15 campsites and picnic area. Adjacent to South Fork of the Yuba River. Good jump-off point into the **back-country.**

(continued)

19...Hampshire Rocks Campground. Take the Rainbow Road exit Off I #80. The campground is located off the frontage road on the freeway. Vault toilets. 31 campsites. Piped water. Trailer units available. Limited parking. Adjacent to South Fork River. Amenities near-by. Back-country has history of placer mining in its creeks. Good site for jumping off into the back-country for recreational gold prospecting. **Do work in research for reservoir and fresh water boundary lines before recreation gold mining work is done.** Contact Agencies for information and availability; 916-692-1121 or 916-272-1375.

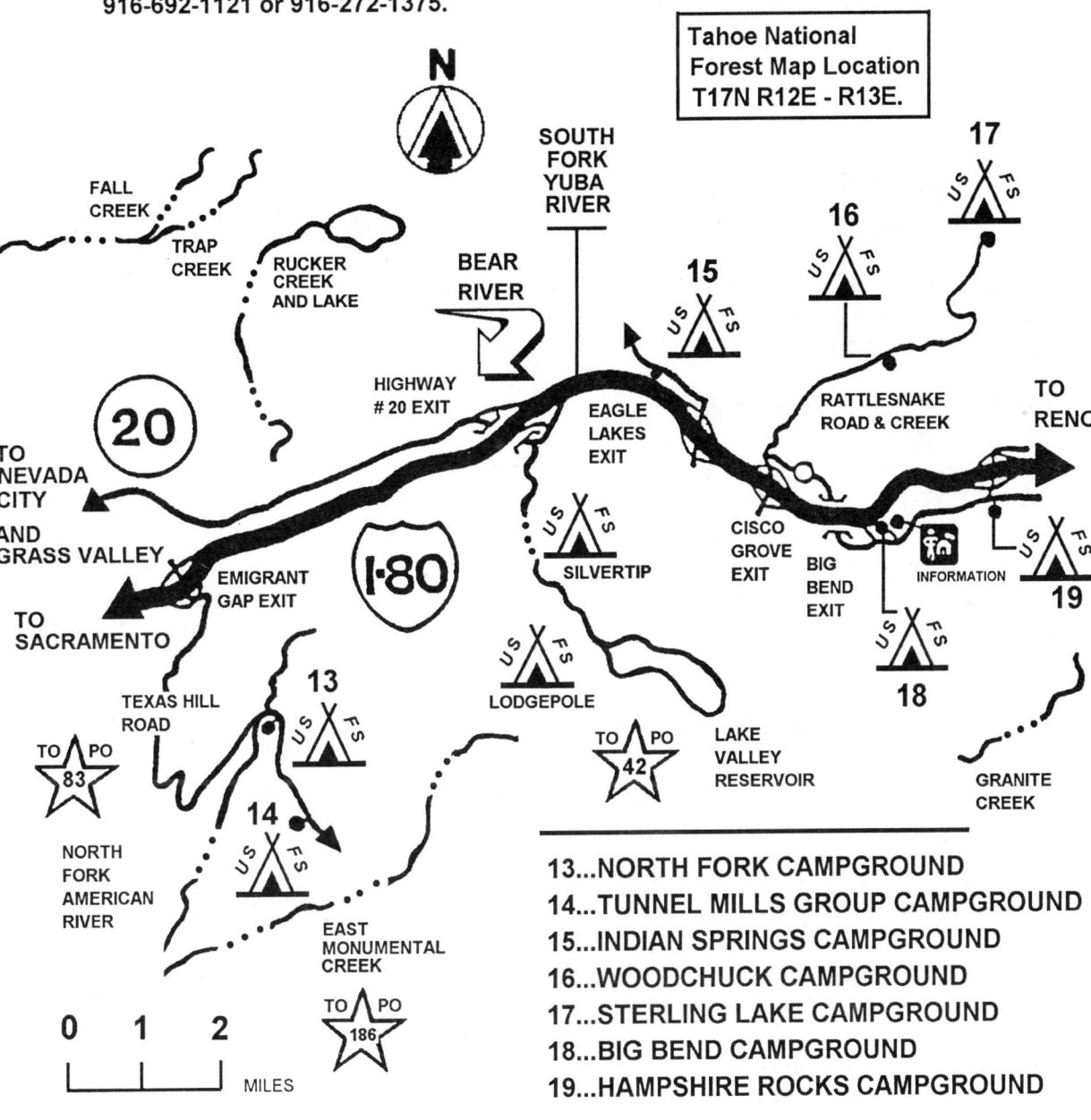

13...NORTH FORK CAMPGROUND
14...TUNNEL MILLS GROUP CAMPGROUND
15...INDIAN SPRINGS CAMPGROUND
16...WOODCHUCK CAMPGROUND
17...STERLING LAKE CAMPGROUND
18...BIG BEND CAMPGROUND
19...HAMPSHIRE ROCKS CAMPGROUND

BULLARDS BAR RESERVOIR - TAHOE NATIONAL FOREST

Bullards Bar Reservoir Area is located on the North Yuba River. It may be reached by traveling Hwy. # 49 to a point 2.5 miles south of Camptonville and then proceeding on a signed country road 4 miles west to the reservoir. From Marysville to Bullards Bar Dam, proceed 11 miles east on Hwy. # 20, then travel 21 miles northeast via Browns Valley and Dobbins. April - October seasonal use. Make base campsite for reaching back-country water-runs and gold prospecting. Amenities are available at North San Juan, Camptonville and Dobbins. **Do work in research for reservoir and fresh water boundary lines before recreation gold mining work is done. Contact Agencies for availability and info.; 916-692-1121 or 916-272-1375.**

20...Schoolhouse Campground. Located off the Marysville Road on the south-east side of Bullards Bar Reservoir. 67 campsites with 33 trailer sites. Piped water. Flush toilets.

21...Cottage Creek Boat Ramp. Located 2 miles southwest of Schoolhouse campground, at the west end of the Dam. Vault toilets. Parking.

22...Dark Day Site (day use). Located 5 miles west of Camptonville just off the Marysville Road. 30 picnic sites with tables and stoves. Vault toilets. Piped water. Concrete boat ramp. Parking. Toilets.

23...Burnt Bridge Campground. Located 5 miles north of the Dam on Challenge Road. 30 campsites with 13 sites for trailers. Piped water. Flush toilets. This campground is not always open, check with Forest Ranger at Star Route, Bx 1, Camptonville, CA., 95922. 916-288-3231.

24...Hornswoggle Group Camp. This site is available by reservation only. Contact the Forest Ranger in the above paragraph for information. This site is located near Schoolhouse Campground. Piped water. Flush and vault toilets. Tables, stoves and campfire circles. Central parking. Group camping for 75 persons.

The following campgrounds are accessible by boat only. Good back-country access for recreational prospecting and gold panning. Campsites have floating chemical recirculating toilets. Burnable refuse should be burned and the remainder brought back to deposit in containers at the boat ramp.

25 Garden Point Campground. Boat access only. 16 campsites. Located one mile northwest of the Dark Day boat ramp on the north side of the reservoir. Purify water.

26...Frenchy Point Campground. Boat access only. 8 campsites. Located three miles northwest of the Dark Day boat ramp on the east shore of the reservoir.

27...Madrone Cave Campsite. Located 5 miles northwest of the Dark Day boat ramp on the west shore of the reservoir. Boat access only. Purify all water use.

20...SCHOOLHOUSE CAMPGROUND
21...COTTAGE CREEK BOAT RAMP
22...DARK DAY (Day Use Site)
23...BURNT BRIDGE CAMPGROUND
24...HORNSWOGGLE GROUP CAMP
25...GARDEN POINT CAMPGROUND
26...FRENCHY POINT CAMPGROUND
27...MADRONE COVE CAMPGROUND

TO THE BACK
OF BULLARDS
RESERVOIR AREA
WITH ACCESS MAP

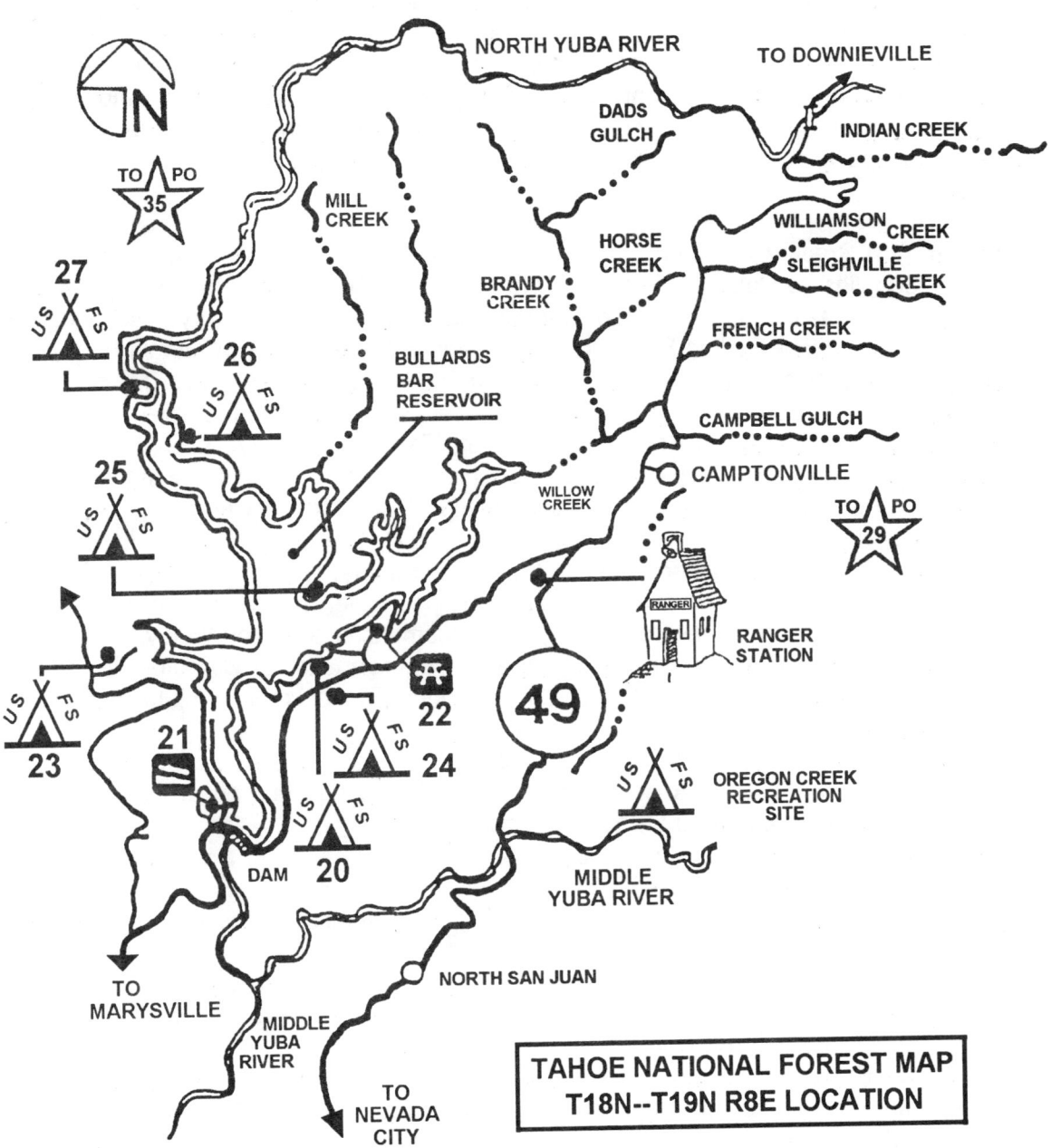

TAHOE NATIONAL FOREST MAP
T18N--T19N R8E LOCATION

49. MIDDLE AND NORTH YUBA RIVERS.

T T19N R8E thrue R11E MAP LOCATION.

49 goes north, crossing the South and Middle Rivers and from Indian Valley through Downieville and Sierra City to following sites are all near Highway # 49 and the Yuba **panning**, nugget sniping, camping, hiking, fishing, son. Overnight camping is permitted only in developed camping areas along the North Yuba River. The back- ater-runs that are seldom visited for recreational gold ng claims and private property within the back-country familiar areas to prospect for gold. A California dredging permit will be required. **Gold panning in California is a Free-Use activity and will not require a permit.** The following campsites are available on a first-come first-serve basis. Please be conscious with the disposing of garbage refuge in these sensitive areas.

28...Oregon Creek (Day use area). 18 picnic units are available. Chemical recirculating toilets. Supplies available at San Juan and Camptonville. **29...Carlton Flat Camping Area.** Lower Carlton Flat is located one-mile northeast of the Highway # 49 bridge at Indian Valley. Upper Carlton Flat is located behind the Indian Valley Outpost on the Cal Ida Road. Portable and Vault Toilets. Stream and river water source. Purify before drinking. Limited supplies at Indian Valley. Undeveloped campsites available.

29...Carlton Flat Camping Area. Lower Carlton Flat is located 1 mile northeast of the #49 bridge at Indian Valley. Upper Carlton Flat is located behind the Indian Valley Outpost on the Cal-Ida Road. Primitive camping. Purify stream water source, Limited supplies at Indian Valley.

30...Fiddle Creek Campground. Located 9 miles northeast of Camptonville on the North Yuba River. 13 campsites available. Vault toilets. No trailer or camper space available. River water source. Purify before drinking. Limited amenities at Indian Valley.

31...Indian Valley Campground. Located 10 miles northeast of Camptonville on the North Yuba River. 17 campsites (8 spaces for trailers). Piped water. Vault toilets. Limited supplies available at Indian Valley.

32...Rocky Rest Camping Area. Located 10 miles northeast of Camptonville on the North Yuba River. Undeveloped camping area. Portable toilets. River water source. Purify before drinking. Limited amenities at Indian Valley.

33...Convict Flat Picnic Site, (Day use only). Located 12 miles northeast of Camptonville on the North Yuba River. 3 picnic sites with tables but no cooking facilities. River water source. Purify water before drinking. Vault toilets. Space for trailer parking; day use only.

34...Ramshorn Campground. Located 15 miles northeast of Camptonville off of Highway # 49. With 16 campsites (7 with space for trailers). Piped water. Vault toilets. Supplies available at Downieville.

35...Indian Rock Picnic Ground. Located 15 miles northeast of Camptonville on the North Yuba River, across from the Ramshorn Campground. With 3 picnic sites accommodating tables and stoves. River water source. Purify before drinking. Vault toilets.

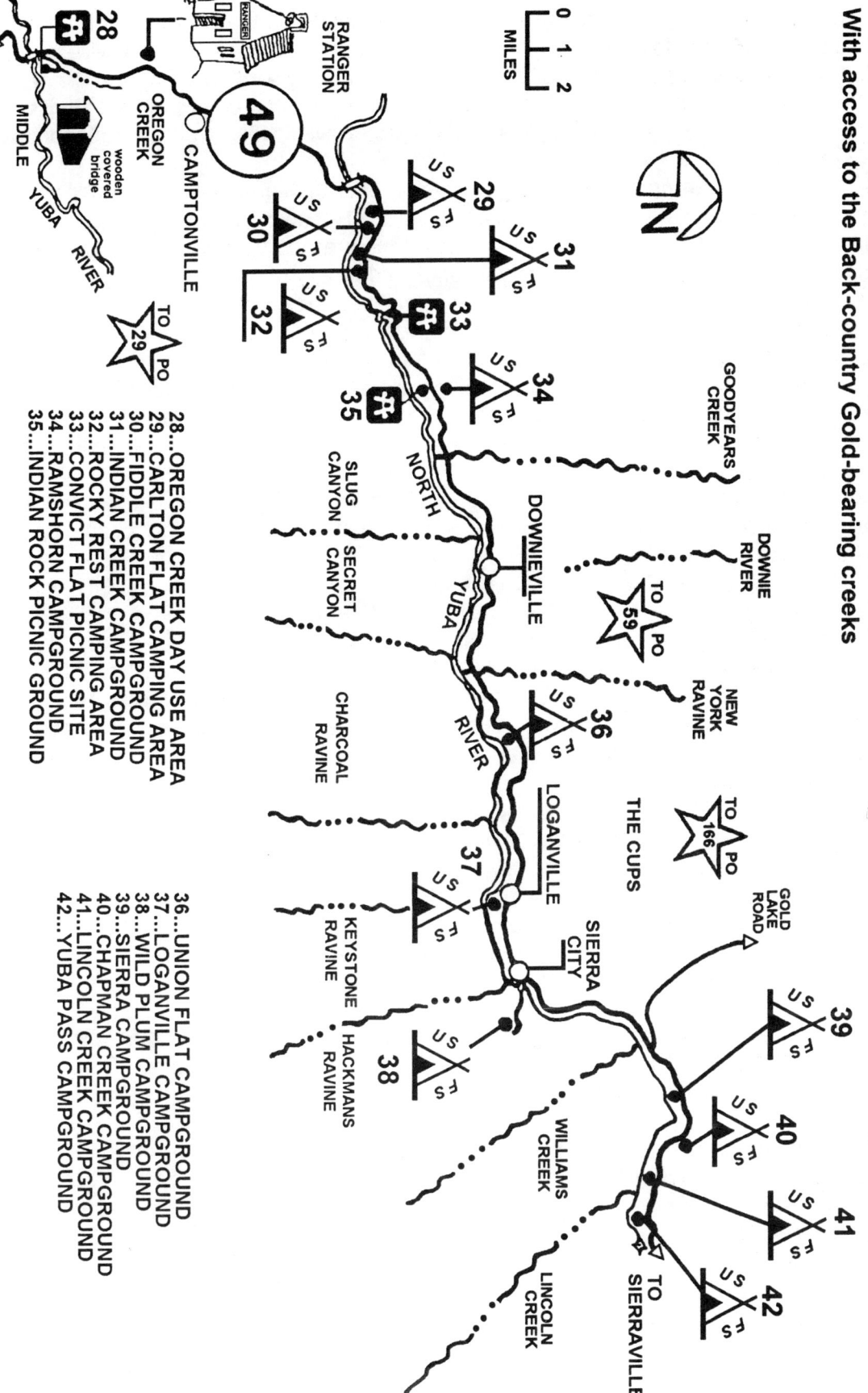

36...Union Flat Campground. 6 miles east of Downieville on the North Yuba River with 12 campsites, 6 sites for trailers, piped water, vault toilets, supplies at Downieville.

37...Loganville Campground. Located 9 miles northeast of Sierra City on the North Yuba River. Nine campsites are available. River water source. Purify water before drinking. Vault toilets. Limited supplies at Bassetts, 4 miles west. Not recommended for larger trailers.

38...Wild Plum Campground. Located one-mile east of Sierra City on the Wild Plum Road along Haypress Creek. With 47 campsites (21 spaces for trailers). Piped water. Vault toilets. Supplies available at Sierra City. This area borders on the ancient belt of Calaveras Schist and Quartzite influence composed mostly of quartz porphyry and meta-sedimentary rocks, pyrite and minor galena; much coarse gold in the past has been recovered leaving remnants of itself for today's gold searcher to hustle for.

39...Sierra Campground. Located 7 miles northeast of Sierra City on the North Yuba River. With 16 campsites (7 have space for trailers). River water source. Purify before drinking. Vault toilets. Supplies available at Bassetts, 2 miles west. Several high-grade surface pockets were recovered near Sierra City in 1850. Hydraulic open-pit mining and surface drifting underwent considerable productivity as evidenced by the ancient scars near Sierra Buttes. It is believed that beneath the present overburden there are in staggered locations those same nugget deposit formations that once laid openly on the surface, such as where the ancient "Tin-cup" deposits exist west of Sierra City.

40...Chapman Creek Campground. Located 8 miles northeast of Sierra City on the North Yuba River. With 29 campsites available with 14 spaces for trailers. Vault toilets. Limited amenities at Bassetts, 3 miles west. In the old days the miner lost a great deal of their gold when they poured mercury into the riffle slots in an attempt to quickly recover the gold occurrences. The old timers lost the gold by not knowing that when they filled those riffle slots brimming over with mercury, the lighter nuggets, spongy gold and the float particles would glide over the mercury and fall back into the Tertiary gravel. These almagamated gold nuggets exist in the water-runs near-by and in the back-country and should be watched for as shiny silvery coated gold nugget objects.

41...Lincoln Creek Campground. Located 2 miles west of Sierra City on the North Yuba River. 15 campsites available during open season. Space for trailers. River water source. Purify before drinking.

42...Yuba Pass Campground. Located 11 miles east of Sierra City and 11 miles west of Sierraville at Yuba Pass Summit. 20 campsites available. Piped water. Vault toilets. Space for small trailers. Supplies at Sattley, 7 miles to the west.

All of the above 42 campsites have been set aside as **FREE-USE MINERAL WITHDRAWAL SITES** where recreational gold prospecting can be pursued. Staking a mineral claim in those designated Free-Use areas is prohibited.

Gold panning, nugget sniping, sluice-rocker, vacuum - suction dredging are activities allowed within the Free-Use areas. Vacuum - dredging requires a California Dredging permit. Gold panning in California is a Free-Use activity and requires no permit.

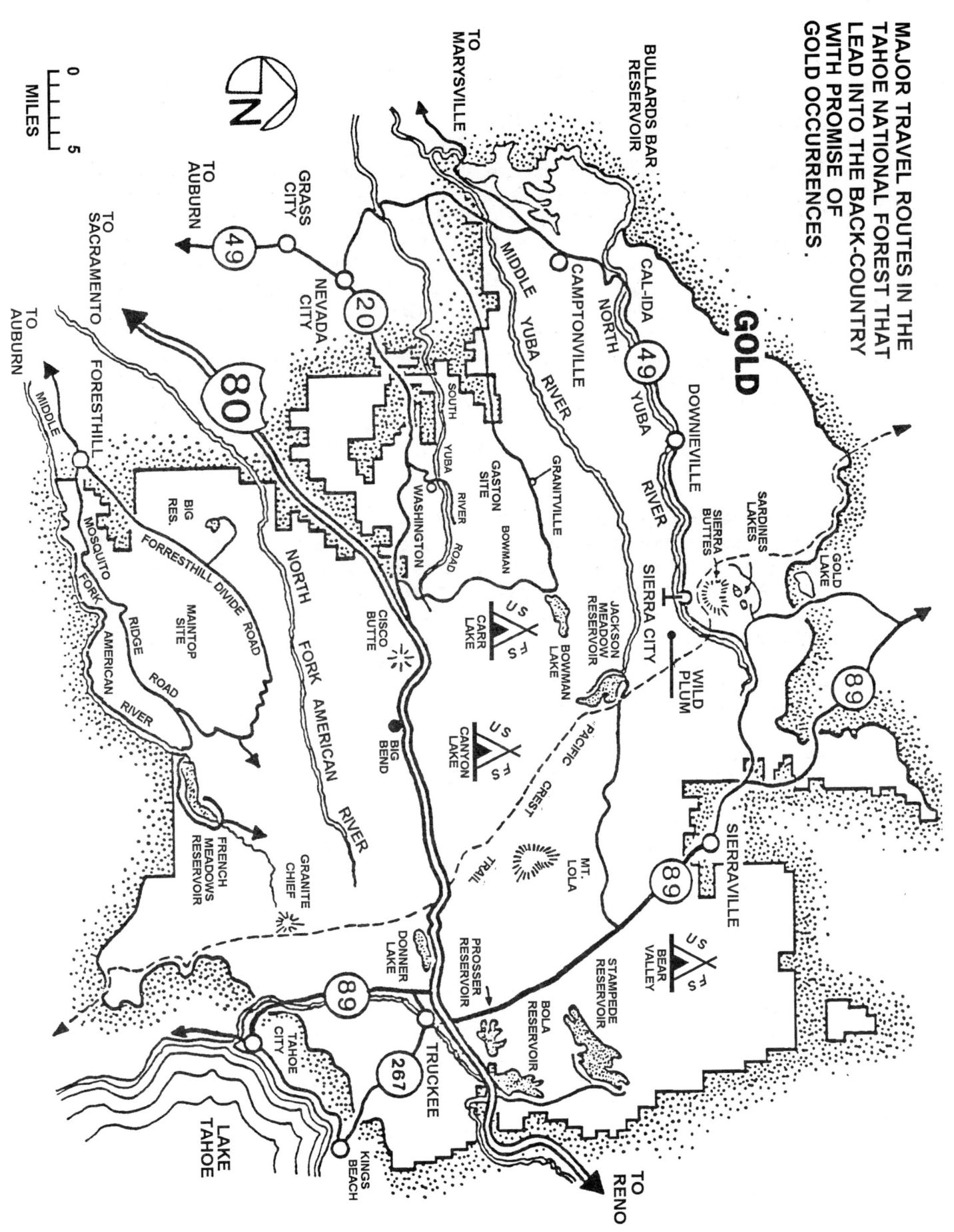

SIERRA BUTTES AND GOLD LAKE HIGHWAY.
TAHOE NATIONAL FOREST,
T20N - T21N R12E MAP LOCATION.

This back-country area of high mountain lakes, water-runs and streams is dominated by the massive and brooding Sierra Buttes. This spectacular land may be reached by following Highway # 49 north and east from Sierra City 4.5 miles to the Gold Lake Highway junction at Bassetts. It may also be reached from Sierra Valley by following Highway # 49 westward approximately 13 miles from its junction with Highway # 89 at Sattley to the Gold Lake Highway. Activities include fishing, swimming, boating, camping, picnicking, hiking and hunting in season. Pacific Crest Hiking Trail is accessible from this area where access to the back-country is available. The back-country offers undiscovered gold occurrences. Supplies and amenities are available at Sierra City and Bassetts. Open season is from June thue October. *Camping is not permitted within one-quarter mile of Gold Lake Highway (from Highway # 49 to Sardine Lake Road), Sardine Lake Road (from Gold Lake Highway to Sardine Lake), Packer Lake Road (from Sardine Lake Road to Packer Lake), and around Salmon Lake because of the heavy visitor use.*

44...Sardine Campground. Located 1.5 miles north of Bassetts on the Gold Lake Highway and 1/2 mile southwest on the Sardine Lake Road. 29 campsites with 15 spaces for trailers are available on a first-come first-served basis. Piped water. Vault toilets. Adjacent to the *back-country gold occurrences.*

45...Salmon Creek Campground. Located 2 miles north of Bassetts on the Gold Lake Highway near Salmon Creek. 32 campsites with 7 spaces for trailers. Piped water. Vault toilets. Access to the un-discovered *back-country gold particles.*

46...Snag Lake Campground. Located 5 miles north of Bassetts on the Gold Lake Road adjacent to Snag Lake. 16 undeveloped campsites with space for small trailers. Lake water source. Purify before drinking. Vault toilets.

47...Packer Lake Picnic Ground. Day use only. Located 3 miles west of the Gold Lake Highway on the Packer Lake Road. 3 picnic sites with tables and stoves. Lake water source. Purify before drinking. Supplies at Packer Lake Lodge.

48...Sand Point Picnic Ground. Day use only. Located just off of the Gold Lake Highway on the Sardine Lake Road near the Sardine Campground. 16 picnic sites with tables and stoves. Pets are to be contained.

49...Berger Campground. Located 2 miles west of the Gold Lake Highway on the Packer Lake Road. 10 undeveloped campsites. Stream water source, purify before drinking water. Space for small trailers. Vault toilets.

50...Diablo Camping Area. Located 1.5 miles west of Gold Lake Highway on Packer Road. Undeveloped Campsites. Room for trailer and tent camping. Stream water source, purify drinking water. Portable toilets.

51...Packsaddle Camping Area. Located 2.5 miles west of Gold Lake Highway on the Packer Lake Road. Undeveloped campsites. Room for trailers and tent camping. Stream water source, purify before drinking. Portable toilets. Horses, and

pack saddle stock permitted. Good jumping off point into the *back-country* for *gold occurrence* prospecting where possibilities are favorable.

TO THE BACK-COUNTRY OF SIERRA BUTTES *AND THE GOLD LAKE AREA OFFERING GOLD OCCURRENCES IN THE WATER-RUNS*

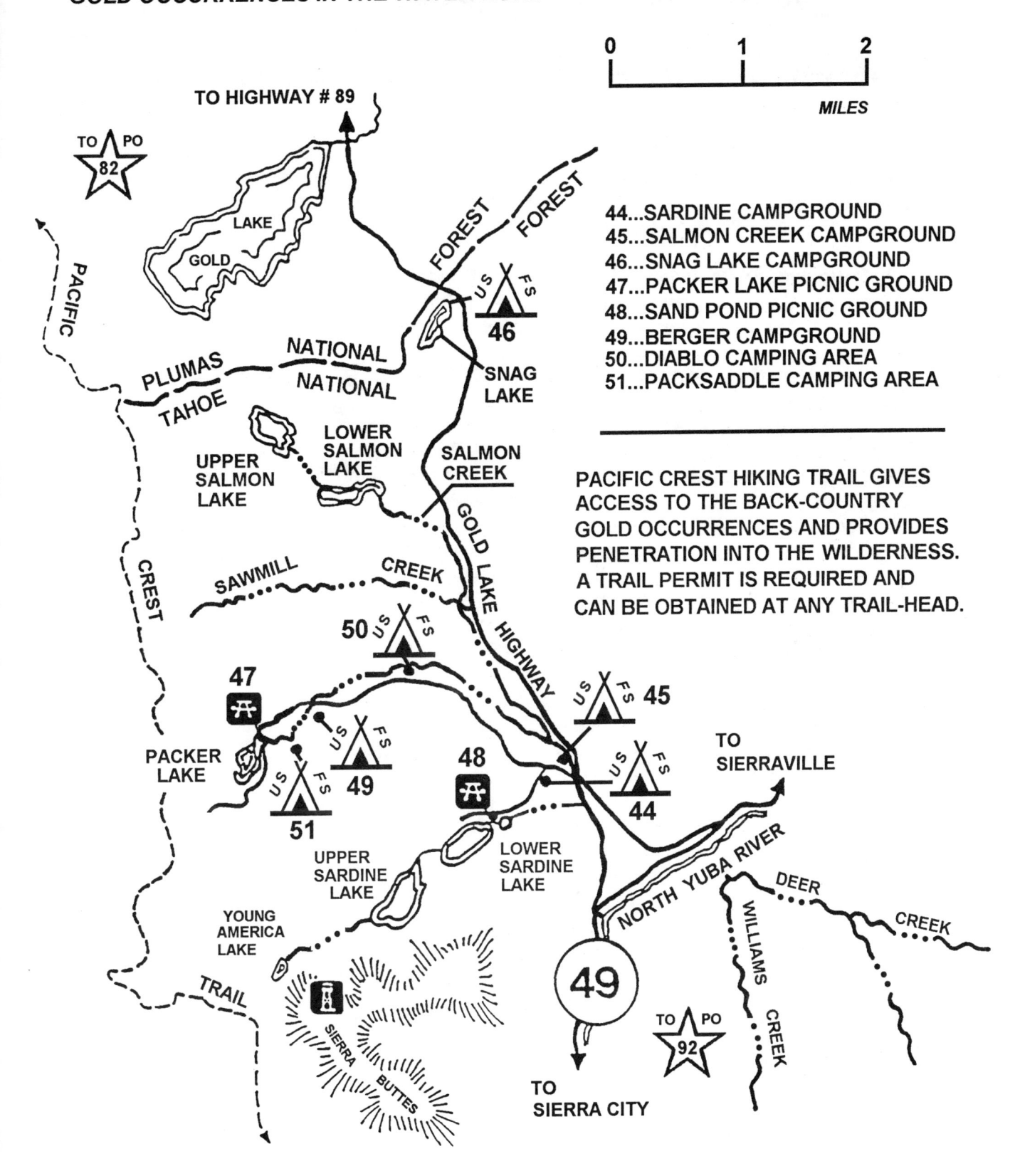

TO HIGHWAY #89

TO PO 82

44...SARDINE CAMPGROUND
45...SALMON CREEK CAMPGROUND
46...SNAG LAKE CAMPGROUND
47...PACKER LAKE PICNIC GROUND
48...SAND POND PICNIC GROUND
49...BERGER CAMPGROUND
50...DIABLO CAMPING AREA
51...PACKSADDLE CAMPING AREA

PACIFIC CREST HIKING TRAIL GIVES ACCESS TO THE BACK-COUNTRY GOLD OCCURRENCES AND PROVIDES PENETRATION INTO THE WILDERNESS. A TRAIL PERMIT IS REQUIRED AND CAN BE OBTAINED AT ANY TRAIL-HEAD.

GOLD LAKE

PLUMAS NATIONAL FOREST
TAHOE NATIONAL FOREST

SNAG LAKE

UPPER SALMON LAKE
LOWER SALMON LAKE

PACIFIC CREST TRAIL

SAWMILL CREEK

GOLD LAKE HIGHWAY

PACKER LAKE

UPPER SARDINE LAKE
LOWER SARDINE LAKE

YOUNG AMERICA LAKE

SIERRA BUTTES

NORTH YUBA RIVER
WILLIAMS CREEK
DEER CREEK

TO SIERRAVILLE
TO SIERRA CITY

TO PO 92

49

32

HIGHWAY # 89 (Sierraville To Truckee).
Access into the back-country with access map.
TAHOE NATIONAL FOREST MAP LOCATION,
T17N - T18N R16E.

This portion of Highway # 89 extends from Sierraville and the National Forest boundary 24 miles south to Truckee and Interstate # 80. These sites are adjacent to Highway # 89 unless stated otherwise. Activities include fishing, camping, hiking, swimming, picnicking and hunting in season. Supplies are available at Sierraville or Truckee. Normal season use is June through October. Overnight camping is permitted only in developed campgrounds or designated sites. *Good jumping off point for recreational gold prospecting into the back-country.*

52...Cottonwood Campground. Located 4.5 miles southeast of Sierraville on Cottonwood Creek. 49 campsites with 28 trailer spaces. Piped water. Vault toilets.
53...Cold Creek Campground. Located 5 miles southeast of Sierraville on Cold Creek. 13 campsites with spaces for small trailers. Piped water. Vault toilets.
54...Bear Valley Campground. Located 8 miles east of Sierraville on the Lemon Canyon Road; (not recommended for trailer use).
55...Upper Little Truckee Campground. Located 9 miles southeast of Sierraville on the Little Truckee River. 26 campsites + trailer space. Piped water. Vault toilets.
56...Lower Little Truckee Campground. Located 9 miles southeast of Sierraville on the Little Truckee River. 15 campsites + trailer space. Piped water. Vault toilets.
57...Sagehen Campground. Located 9 miles north of Truckee and 2 miles west of Highway # 89. Turn left at the Sagehen Summit turnoff. 10 undeveloped campsites. Vault toilets. Creek water source, purify before drinking. Pack-in, pack-out refuge.
58...Annie McCloud Campground. Located 5 miles northeast of Truckee on the northeast shore of Prosser Reservoir. From Highway # 89 turn right on Prosser Dam Road, 1/2 mile north of I-80, and continue 4.5 miles. The campground is just past the dam on the left. 10 campsites with space for small trailers. Reservoir water source, purify before using. Vault toilets, (across the road from the campground).
59...Lakeside Campground. Located 4 miles north of Truckee on Highway # 89 on the northwest shore of Prosser Reservoir. 30 undesignated campsites, with space for trailers. Reservoir water source, purify before using. Vault toilets. Boating.
60...Prosser Campground. Located 3 miles north of Truckee on Highway # 89 on the west shore peninsula of Prosser Reservoir. 29 campsites. Piped water. Vault toilets. Concrete boat launching ramp near-by.
61...Prosser Group Campground. Located 3 miles north of Truckee on Highway # 89, on the west shore peninsula of Prosser Reservoir. This group campground is available by reservation only. For information & reservations contact; Truckee Ranger District, Box 399, Truckee, CA., 95734. 916-587-3558.
62...Donner Camp Historic & Picnic Site, (day use only). Located 2.5 miles north of Truckee on Highway # 89, on the southwest shore of Prosser Reservoir. Vault toilets. 4 picnic sites with tables and fire rings. No water, dry camp site. *The above campgrounds are good jumping off points for the back-country to do gold-search prospecting and nugget sniping.*

ACCESS INTO THE BACK-COUNTRY WITH HIGHWAY # 89 VIA SIERRAVILLE AND TRUCKEE. TAHOE NATIONAL FOREST MAP LOCATION T17N - T18N R16E.

Quote, "........one hundred billion dollars in gold is laying waiting to be extracted from the gravel in the back-country by an economical process that would not upset the ecology". Several millions of years in the past, the Tertiary ages in geological history hosted hundreds of ancient rivers sloshing with gold particles which ran their course over much older terraces of land also saturated with gold occurrences, that later became known as the High Sierras through slow moving forces that had thrusted the northern Sierra Nevada influence into spectacular peaks exposing over 600 miles in length and 100 miles plus in width, with gold bearing double layered Tertiary gravel channels. In the years which followed, much of the terraced land underwent a roller-coaster movement, shifting new and younger rivers to lay on top of those older river beds and slipping vast amounts of gold layered gravels to newer Quaternary positions. Rivers were influenced to flow into new courses through the ancient land, leaving deep canyons filled with immense layers of gold debris in ancient riffles.

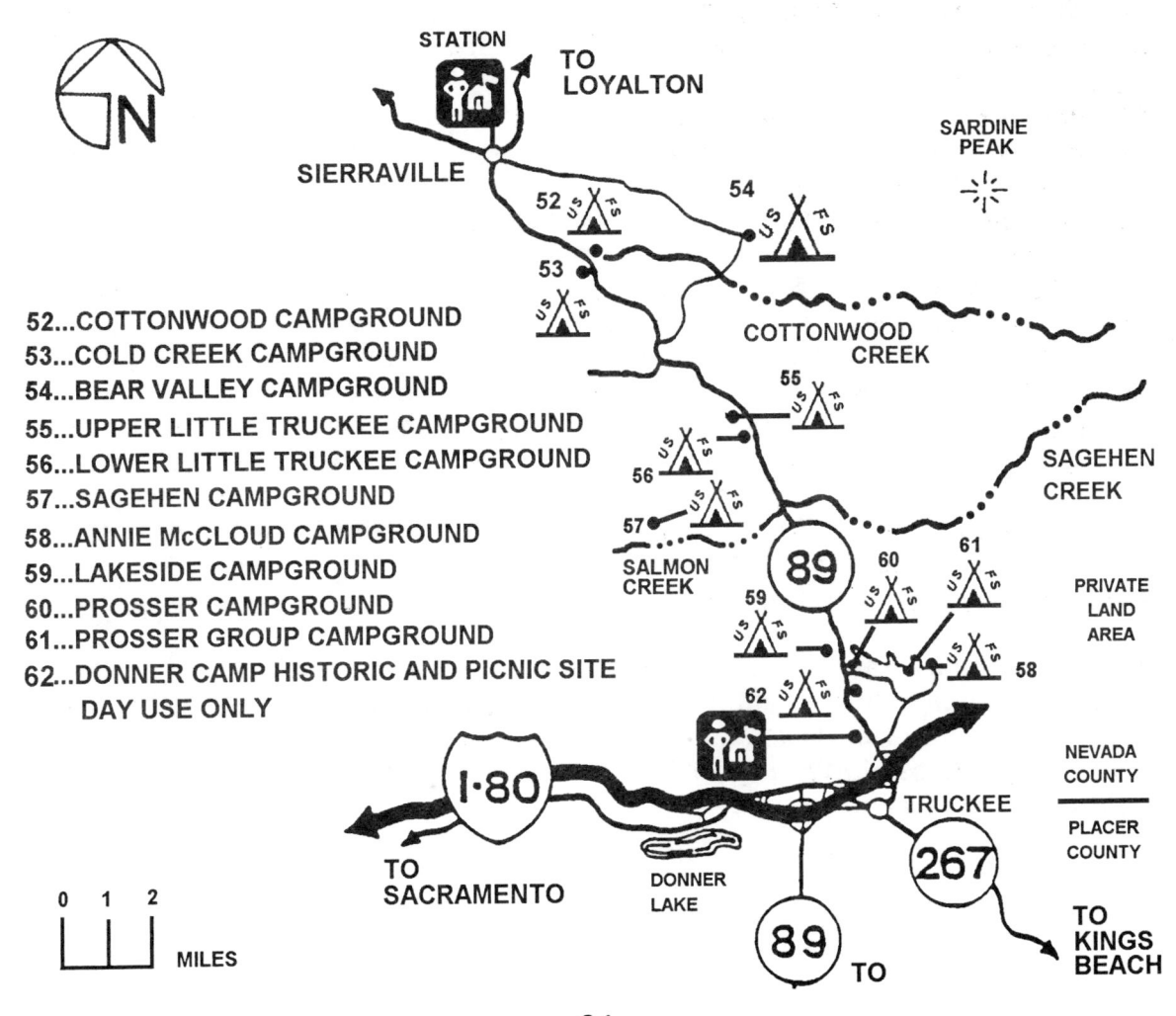

52...COTTONWOOD CAMPGROUND
53...COLD CREEK CAMPGROUND
54...BEAR VALLEY CAMPGROUND
55...UPPER LITTLE TRUCKEE CAMPGROUND
56...LOWER LITTLE TRUCKEE CAMPGROUND
57...SAGEHEN CAMPGROUND
58...ANNIE McCLOUD CAMPGROUND
59...LAKESIDE CAMPGROUND
60...PROSSER CAMPGROUND
61...PROSSER GROUP CAMPGROUND
62...DONNER CAMP HISTORIC AND PICNIC SITE
    DAY USE ONLY

JACKSON MEADOW RESERVOIR AREA WITH MAP LOCATION FOR TAHOE NATIONAL FOREST AT T19N R13E.

Proceed 17.5 miles north of Truckee on Highway # 89 and 16 miles west on the Forest Road # 07 to Jackson Meadow Reservoir. A secondary access is by way of the Bowman Road. A Forest Service Station is located at the southwest side of the lake, with a self-guided nature trail near-by. Supplies are available in Truckee and Sierraville. Seasonal use is June thrue October. Overnight camping is permitted only in developed campgrounds. A recreational vehicle refuge dump is across from the information turn-out. *The following sites make good base camps for jumping off into the back-country in search of the gold occurrences.*

63...PASS CREEK CAMPGROUND. Located on the northeastern shore of Jackson Meadow Reservoir. With 30 campsites and 15 of those have spaces for trailers. Piped water. Flush toilets. Swimming. Boat launching ramp.

64...JACKSON POINT CAMPGROUND. Boat access only. Located on the south shore of the reservoir 1/2 mile southwest of the Pass Creek boat ramp. With 10 campsites. Reservoir water source. Purify before using. Vault toilets.

65...EAST MEADOW CAMPGROUND. Located on the northeast shore of Jackson Meadow Reservoir. With 46 campsites and 26 of those have spaces for trailers. Piped water. Flush toilets.

66...WOODCAMP CAMPGROUND. Located on the southwest shore of Jackson Meadow Reservoir. With 20 campsites and 10 of those have spaces for trailers. Piped water. Flush toilets.

67...FINDLEY CAMPGROUND. Located on the west side of Jackson Meadow Reservoir. With 12-single-family campsites and three 2-family campsites. Trailer space available. Piped water. Flush toilets.

68...FIR TOP CAMPGROUND. Located on the west side of Jackson Meadow Reservoir. With 11 single-family campsites one 2-family campsite. Piped water. Flush toilets.

69...WOODCAMP PICNIC SITE. (Day use only). Located on the southwest shore of Jackson Meadow Reservoir, adjacent to the Woodcamp Campgrounds. With 15 picnic sites with tables and stoves. Piped water. Vault toilets. Swimming. Dressing room facilities. Boat launching ramp below Woodcamp Campground.

70...SILVER TIP GROUP CAMPGROUND. Located adjacent to the Woodcamp Campground on the southwest shore of Jackson Meadow Reservoir. This site available by reservation only for groups. For reservations contact; Sierraville Ranger Station, P.O. Box 95, Sierraville, CA., 96126.   916-994-3401. Two 25-person group campsites. Tables, stoves, campfire circle. Piped water. Vault toilets. Central parking. Swimming beach at Woodcamp Picnic Site near-by.

The back-country saw much test hole digging, ditch channeling and openpit mining in search of the ancient river placer gold amidst the Tertiary gravel stream beds. The overgrowth, vegetation and shrub growth has almost obscured the evidence of those old timers work. Be extremely cautious of the dangers of the out-back open-pits and holes.

71...ASPEN PICNIC SITE. Day use only. Located on the northeast shore of Jackson Meadow Reservoir. 15 picnic sites with tables and stoves. Piped water. Vault & flush toilets. Concrete boat ramp at Pass Creek Campground. Swimming beach with dressing room facilities.

72...ASPEN GROUP CAMP. Located adjacent to the Aspen Picnic Site on the northeast shore of the Jackson Meadow Reservoir. This site is by reservation only. Contact; Sierraville Ranger District, P. O. Box 95, Sierraville, CA., 96126. 916-994-3401. Two 25 person group campsites, one 50-person group campsite. Tables, stoves, campfire circle. Central parking. Piped water. Vault toilets. Swimming beach and boat ramp near-by.

73...WOODCAMP CREEK INTERPRETIVE TRAIL. A self-guided nature trail is located east of the Forest Service Station on the southwest corner of the Jackson Meadow Reservoir. This trail is about 1/2 mile in length and explains some of the natural features found in the area. *(These campsites make good jumping off points for entering the back-country to do gold prospecting from).*

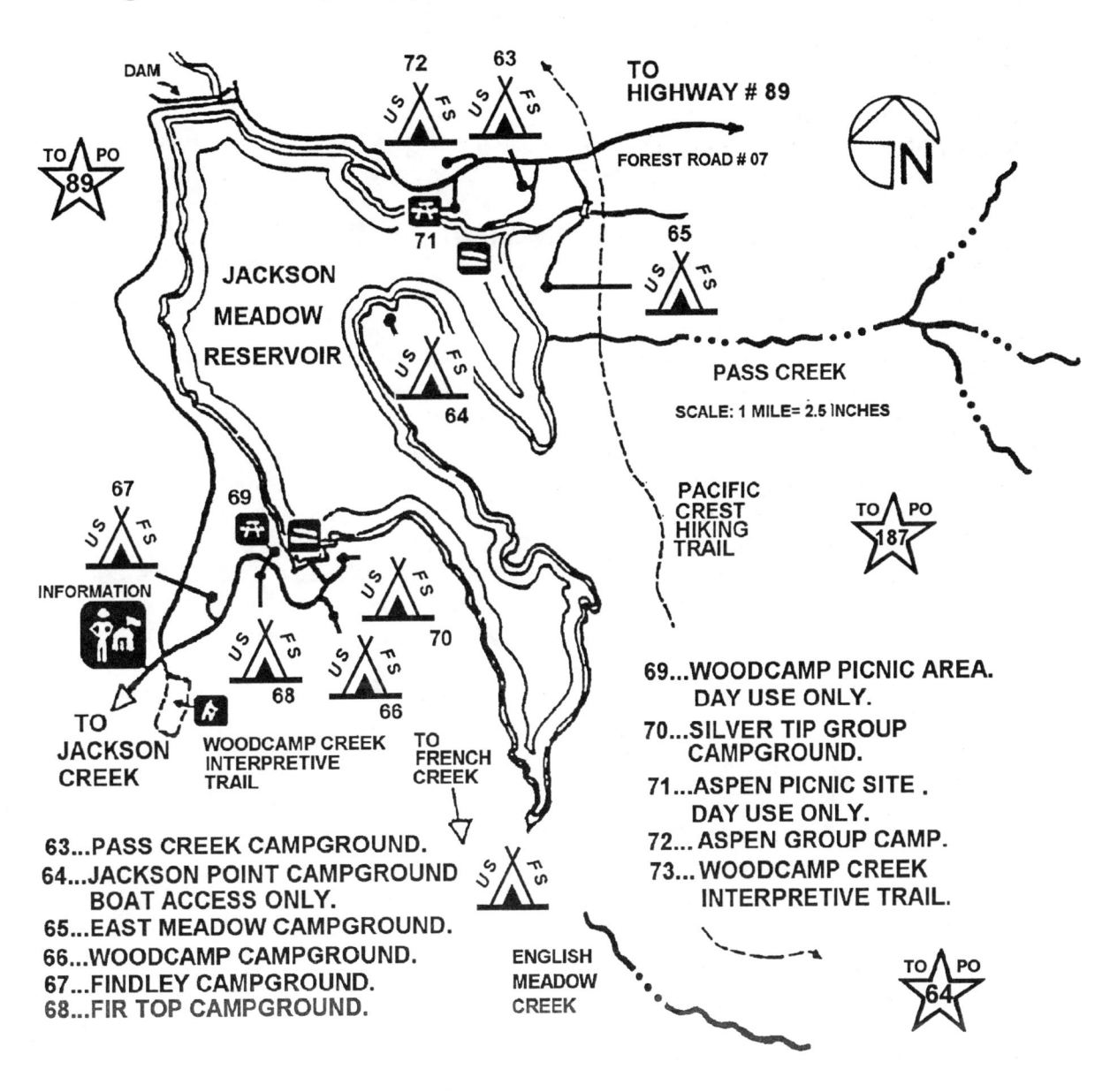

63...PASS CREEK CAMPGROUND.
64...JACKSON POINT CAMPGROUND BOAT ACCESS ONLY.
65...EAST MEADOW CAMPGROUND.
66...WOODCAMP CAMPGROUND.
67...FINDLEY CAMPGROUND.
68...FIR TOP CAMPGROUND.
69...WOODCAMP PICNIC AREA. DAY USE ONLY.
70...SILVER TIP GROUP CAMPGROUND.
71...ASPEN PICNIC SITE. DAY USE ONLY.
72...ASPEN GROUP CAMP.
73...WOODCAMP CREEK INTERPRETIVE TRAIL.

STAMPEDE AND BOCA RESERVOIR AREAS. TAHOE NATIONAL FOREST MAP LOCATION. T18N - T19N R17E.

Both reservoirs are located on the Little Truckee River. Take the Boca-Hirschdale off-ramp off I-80, seven miles east of Truckee. Overnight camping permitted only in developed campgrounds or designated camping areas. Use season June thru October. Truckee is nearest location for amenities and supplies. *Good area to make base camp for jumping off into the back-country to search for gold particles and engage in recreational prospecting. Check with local Forest Ranger for information about the fresh water boundary lines before serious mining has been started.*

74...BOCA CAMPGROUND. Located on the southwest shore of Boca Reservoir. 20 campsites with limited space for trailers. Lake water source. Purify before using. Portable toilets. Concrete boat ramp north of campground.

75...BOCA REST CAMPGROUND. Located on the northeast shore of Boca Reservoir, 2 miles north of the Dam. 25 tent or trailer campsites. Piped water. Vault toilets. Concrete ramp on southwest shore 3 miles away. Hand-off boat launching.

76...BOYINGTON MILL CAMPGROUND. Located on the Little Truckee River, 4 miles north of the Boca Dam. 10 tent or trailer sites. River water source. Purify before using. Vault toilets.

77...STAMPEDE RESERVOIR AREA. EMIGRANT GROUP CAMP. Located on the southeast shore of Stampede Reservoir, 8 miles from the Boca Dam and 1/2 mile west of the Stampede Dam. This group camp is available by reservation only. For information contact; California Land Management, 675 Gilman St., Palo Alto, CA., 94301. Two 25-person campsites. Two 50-person campsites. Tables, fire rings and BBQ's. Firewood scarce. Recirculating toilets. *Trailer refuge dump station.*

78...LOGGER CAMPGROUND. Located on the south side of Stampede Reservoir, one-mile west of Stampede Dam. 252 tent or trailer campsites. Piped water. Vault and chemical recirculating toilets. Several multi-family campsites. Boat launch near.

79...CAPTAIN ROBERTS BOAT RAMP. 2 miles west of Stampede Dam. Vault toilets. Multilane concrete boat ramp. Parking area. *Stampede Reservoir Area is Concessionaire Operated. Check with the California Land Management for information at; 415-322-1183. Address is listed above.*

★ ★ ★ ★ ★ ★ ★ ★

The following list naming Hydraulic Open-Pit Mines are just a few of many that can be found throughout California's back-country. These can be found on the Tahoe National Forestry Map by Tier, Range and Section number;

CHALK BLUFF...T16N...R10E...Sec.29.
RED DOG...T16N...R10E...Sec.30.
YOU BET...T16N...R10E...Sec.31.
ROUGH AND READY...T16N...R8E...Sec.24.
LITTLE NORTH COLUMBIA...T18N...R9E...Sec.32.
MALAKOFF DIGGINS...T17N & T18N...R9E & R10E...Sec1 and 36.

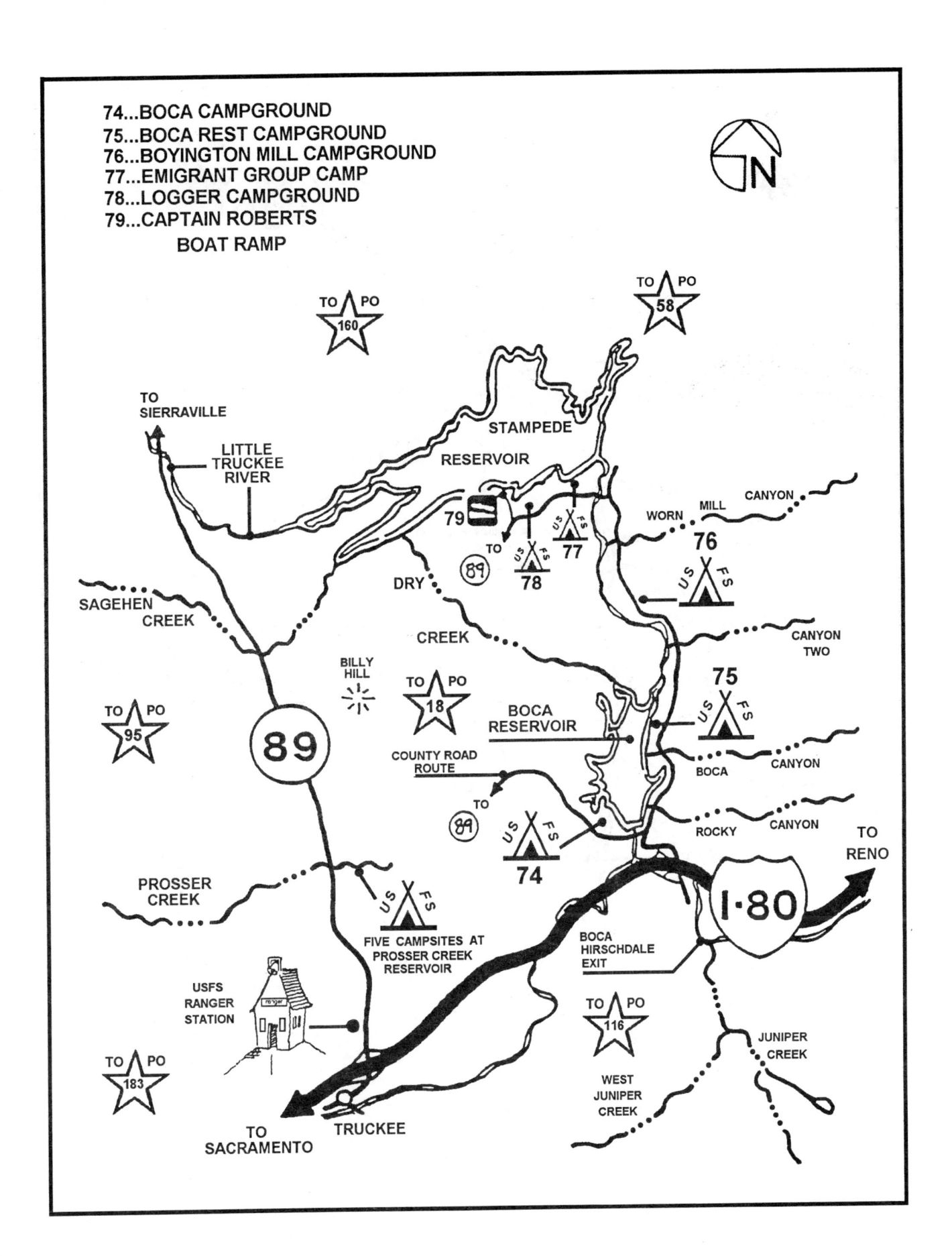

74...BOCA CAMPGROUND
75...BOCA REST CAMPGROUND
76...BOYINGTON MILL CAMPGROUND
77...EMIGRANT GROUP CAMP
78...LOGGER CAMPGROUND
79...CAPTAIN ROBERTS
       BOAT RAMP

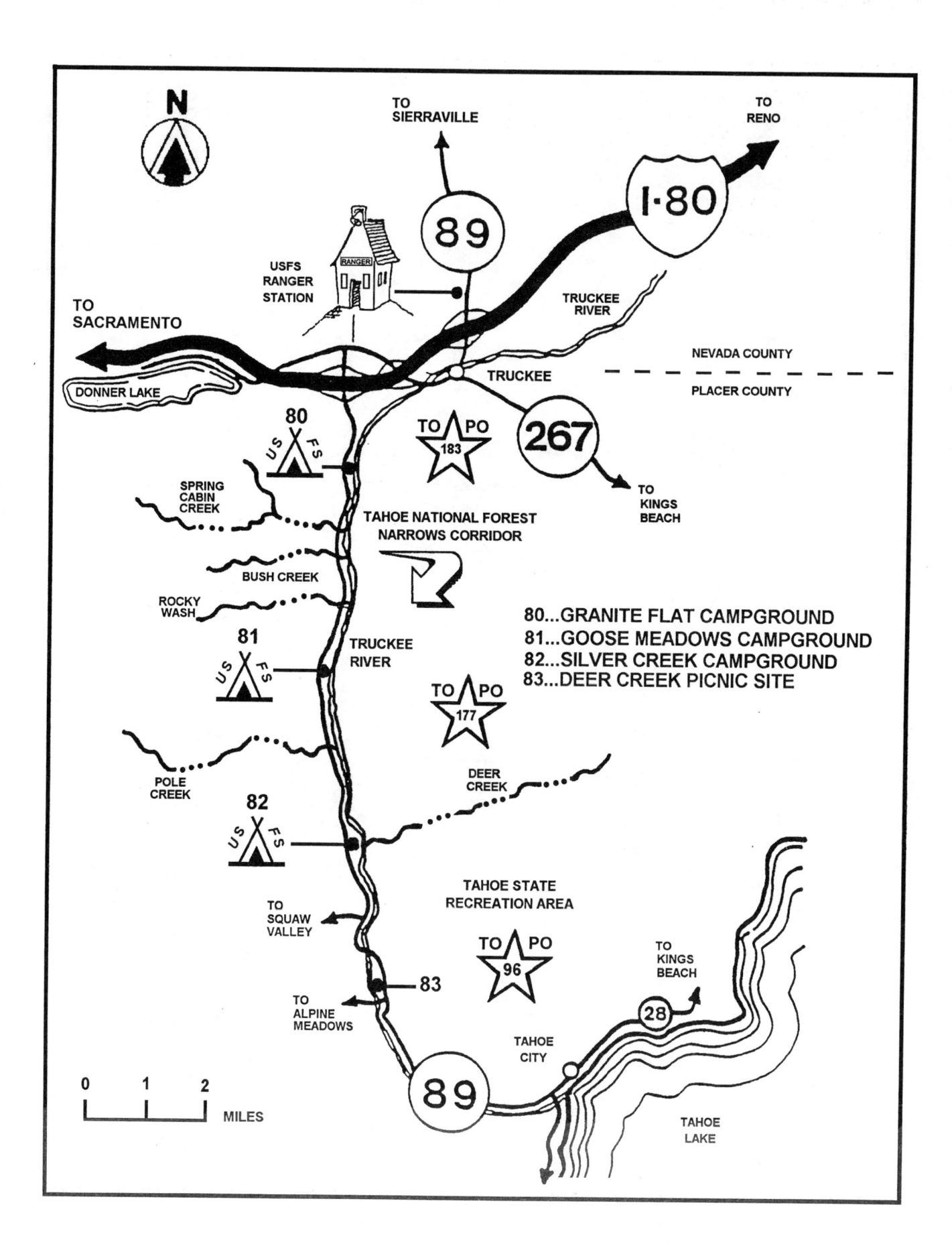

80...GRANITE FLAT CAMPGROUND
81...GOOSE MEADOWS CAMPGROUND
82...SILVER CREEK CAMPGROUND
83...DEER CREEK PICNIC SITE

FORESTHILL DIVIDE--BIG RESERVOIR AND MOSQUITO RIDGE ROAD AREA.
TAHOE NATIONAL FOREST ACCESS MAP T14N / T15N R10E THRUE R12E.

Wherever there had been Hydraulic Mining activities in years past, will indicate to the gold searcher that the surrounding back-country held ancient gold occurrences in the near-by ancient Auriferous gravel beds. Establishing a base camp in the Forestry and BLM campsite gives the gold searcher a close-encounter with the 1850's gold discoveries in areas which the old-timers had to abort their discoveries because of the U.S. Supreme Court decision forbidding Hydraulic Mining.
The area is heavily forested, mountainous with narrow winding, and in some cases just plain graveled roads.
North Fork American River has a history of placer gold in all of its tributaries. Yearly 'n seasonal water run-off flushes the land of gold particles sending these into those adjacent creeks and streams.

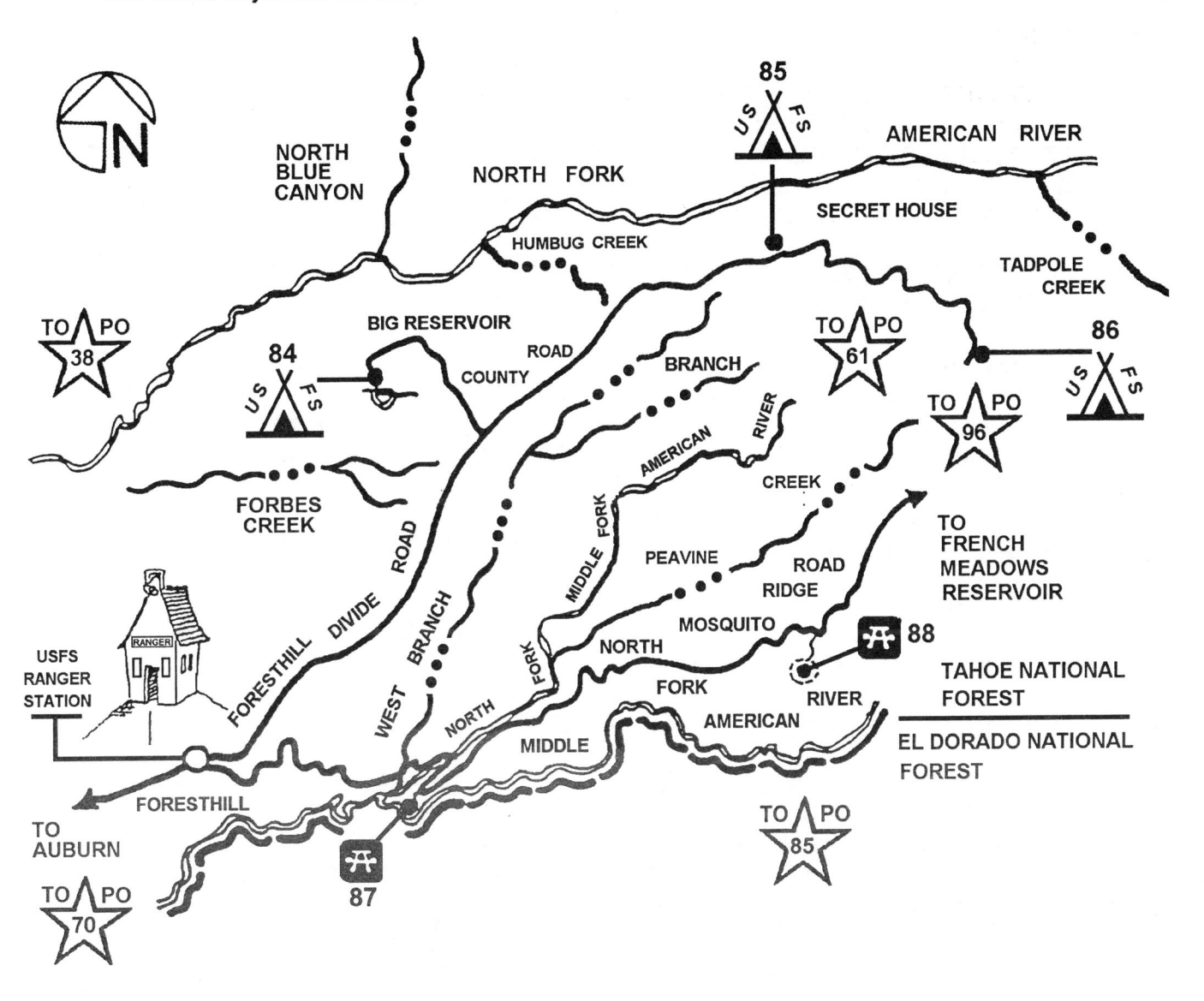

TRUCKEE RIVER CORRIDOR--STATE ROUTE # 89 SOUTH.
TAHOE NATIONAL FOREST MAP LOCATION, T17N - T15N R16E.

There are 3 campgrounds and 1-picnic area on Highway # 89 south of Truckee. All sites are along the Truckee River and are contingent to land under private control. Permission should be obtained before entering upon private property. Nearest location for supplies at Truckee and Tahoe City. Normal season is from July through September. Search deeper into the distant and vacant land for gold occurrences. Past mining production for gold occurrences was a disappointment to the old miners as deeply seated, scattered and stingy veins of gold were discovered.

80...GRANITE FLAT CAMPSITES. Located 1.5-miles south of Truckee on Highway 89. With 75 undesignated campsites. River water source, purify before using. Vault toilets. Ample parking.

81...GOOSE MEADOWS CAMPGROUND. Located 4-miles south of Truckee on Highway # 89. With 25 campsites. River water source, purify before using. Vault toilets. *Limited parking.*

82...SILVER CREEK CAMPGROUND. Located 6-miles south of Truckee on Highway 89. With 29 campsites (19 have trailer space). Piped water . Vault toilets.

83...DEER PARK PICNIC SITE, (day use only). Located 10-miles south of Truckee on the Truckee River . With 5 picnic sites. River water source, purify before using. Vault toilets. *Access not suitable for trailers.*

FORESTHILL DIVIDE--RESERVOIR AREA. TAHOE NATIONAL
FORESTRY LOCATION MAP, T14N / T15N R11E / R13E.

84...BIG RESERVOIR CAMPGROUND. Located 13-miles north of Foresthill. For information for group reservations, contact DeAnza Placer Mining Company, P.O. Box 119, Foresthill, CA., 95631. With 19 campsites. Limited firewood. Piped water. Vault toilets. With 4-picnic sites with table and fire rings. *Limited trailer space.*

85...SECRET HOUSE CAMPGROUND. Located 19-miles northeast of Foresthill on Foresthill Divide Road. Narrow, winding, unpaved mountain road. *Trailers not recommended.* With 2 campsites. Vault toilets. Creek water source, purify before using. Pack-in / Pack-out, (no garbage receptacle).

86...ROBINSON FLAT CAMPGROUND. Located 27-miles northeast of Foresthill on Foresthill Divide Road. Narrow, winding, unpaved mountain road. *Trailers not recommended.* With 6 campsites. Vault toilets. Pack-in / Pack-out, (no garbage cans).

MOSQUITO RIDGE ROAD-FOREST ROAD # 96.

87...RALSTON PICNIC SITE, (day use only). Located 12-miles east of Foresthill on Mosquito Ridge Road # 96 and 1-mile south on Forest Road # 23. Located on North Fork of the American River. *GOLD PANNING* at 5 picnic sites with tables and fire rings. River water source, purify before using. Vault toilets.

88...BIG TREES GROVE. With 3 picnic sites, tables and grills. Piped water. *Use these sites for base camps while searching the back-country* for *gold.*

FRENCH MEADOWS RESERVOIR AREA WITH CAMPSITES.
TAHOE NATIONAL FOREST MAP LOCATION, T15N R14E.

Located 36-miles east of Foresthill by way of Mosquito Ridge Road # 96. This recreation area is within the State Game Refuge and no firearms are permitted. *Recreational gold mining is allowed outside of the Refuge.* Check for fresh water and reservoir boundaries before prospecting.

89...French Meadows Campground and Forest Station.
Located on the south shore of the reservoir with 75 campsites. Piped water. Flush toilets. Concrete boat launching ramp near-by.

90...French Meadows Picnic Area. Day use only. Located on the south shore of the reservoir. Boat ramp. With 7 picnic sites, tables and cooking grills. Campfire circle. Piped water. Flush toilets.

91...Gates Group Camp. Located east side of reservoir. By reservation only. Contact; Foresthill Ranger District, 22830 Foresthill Road, Foresthill, CA., 95631. 916-367-2224. Two 25-person campsites. One 75-person campsite. Campfire circle. Piped water. Vault toilets. Parking.

92...Coyote Group Campground. Located northeast on the shore of the reservoir. By reservation only. Contact the above Ranger address. Three 25 person campsite. One 50-person campsite. Piped water. Toilets.

93...Lewis Campground. Located on the north shore of the reservoir. With 40 campsites. Piped water. Flush toilets. Boat ramp.

94...McGuire Picnic Site. Day use only. Located on the north shore of the reservoir. With 10 picnic sites, tables and cooking grills. Piped water. Flush toilets. Central parking. Beach with swimming facilities.

95...Poppy Campground. Located on the northwest shore of the reservoir. Access by boat or trail only. Approximate one-mile walking distance from McGuire parking lot. With 12 campsites. Vault toilets. River water source, purify before using.

96...Ahart Campground. Located on the middle Fork of the American River. One-mile above the reservoir with 12 campsites on the Mosquito Ridge Road. Vault toilets. River water source, purify before using.

97...Talbot Campground. Located on the Middle Fork of the American River five-miles above the reservoir with 5 campsites. Vault toilets. River water, purify before using. Near trailhead to the Granite Chief area.

The above campsites make excellent base campsites for jumping off into the back-country to do recreational mining in the water-runs. History has experienced hydraulic mining and lode mining in the stream placer gravel.

DELOS TOOLE ©

BACKCOUNTRY ACCESS MAP INTO THE FRENCH MEADOWS RESERVOIR AREA WHERE IT HOLDS A HISTORY OF PLACER GOLD MINING WITH ITS WATER-RUNS GIVING UP GOLD PARTICLES AND GOLD OCCURRENCES.

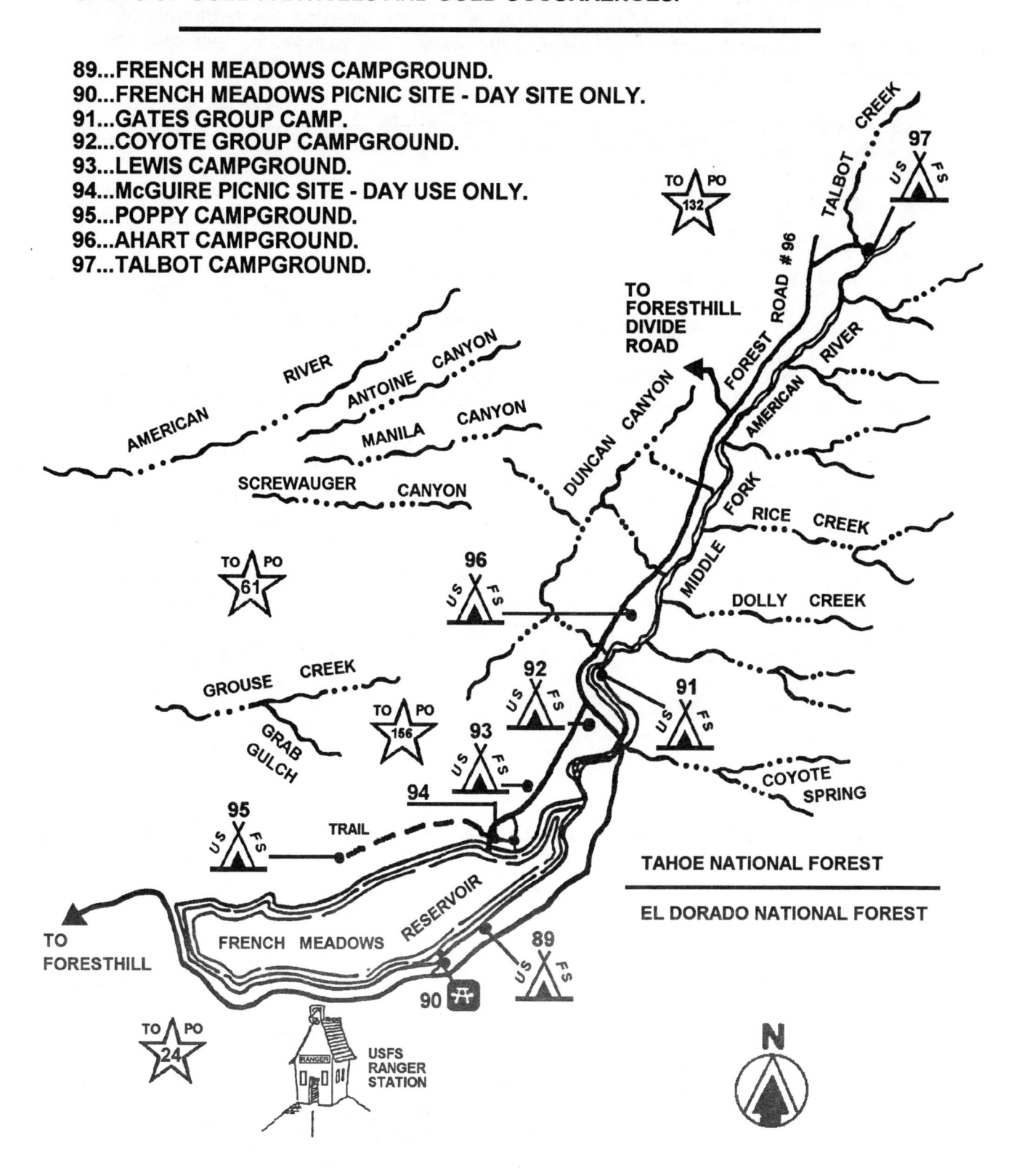

89...FRENCH MEADOWS CAMPGROUND.
90...FRENCH MEADOWS PICNIC SITE - DAY SITE ONLY.
91...GATES GROUP CAMP.
92...COYOTE GROUP CAMPGROUND.
93...LEWIS CAMPGROUND.
94...McGUIRE PICNIC SITE - DAY USE ONLY.
95...POPPY CAMPGROUND.
96...AHART CAMPGROUND.
97...TALBOT CAMPGROUND.

SUGAR PINE RESERVOIR. TAHOE NATIONAL FOREST.

This recreation complex was opened in May of 1985. It contains two group campgrounds, 2-family campsites, boat ramp, hiking trails, picnic area and with swimming beach facilities and near-by is a *trailer dump station.*

Located 15-miles northeast of Foresthill just off of the Sugar Pine Road. Accommodations for the handicapped are indicated by blue signs, but are not solely reserved for the handicapped, (the public has access to these).

98...Forbes Creek Group Campground. Two 50-person group site. Campfire circles. Piped water. Vault toilets. Central parking to accommodate recreation vehicles. This site by reservation only. Contact; Foresthill Ranger District, 22830 Foresthill Road, Foresthill, CA. 95631. 916-367-2224.

99...Giant Gap Campground. With 30 campsites. Piped water. Vault toilets. Fire rings. Campsites will accommodate 30-foot recreational vehicles.

100...Shirttail Creek Campground. Campsites can accommodate 30-foot recreational vehicles. With 30 campsites. Piped water. Vault toilets. Fire rings. Fishing and swimming near-by.

101...Manzanita Picnic Area. Day use only. With 23 picnic sites. Piped water. Vault toilets. Swimming beach. One-mile paved trails along the reservoir shore.

USE CAMPGROUNDS FOR BASE CAMPSITES WHILE ENTERING INTO THE BACK-COUNTRY IN SEARCH FOR GOLD CARRYING STREAMS. AREA HAS A HISTORY OF PLACER MINING. BEFORE BEGINNING SERIOUS GOLD MINING, CHECK WITH AREA FOREST RANGER FOR LOCATION OF BOUNDARY LINES OF FRESH WATER AND RESERVOIR BASINS.

SUGAR PINE RESERVOIR BASIN AREA WITH TAHOE NATIONAL FOREST ACCESS MAP TO THE BACK-COUNTRY T15N R11E LOCATION.

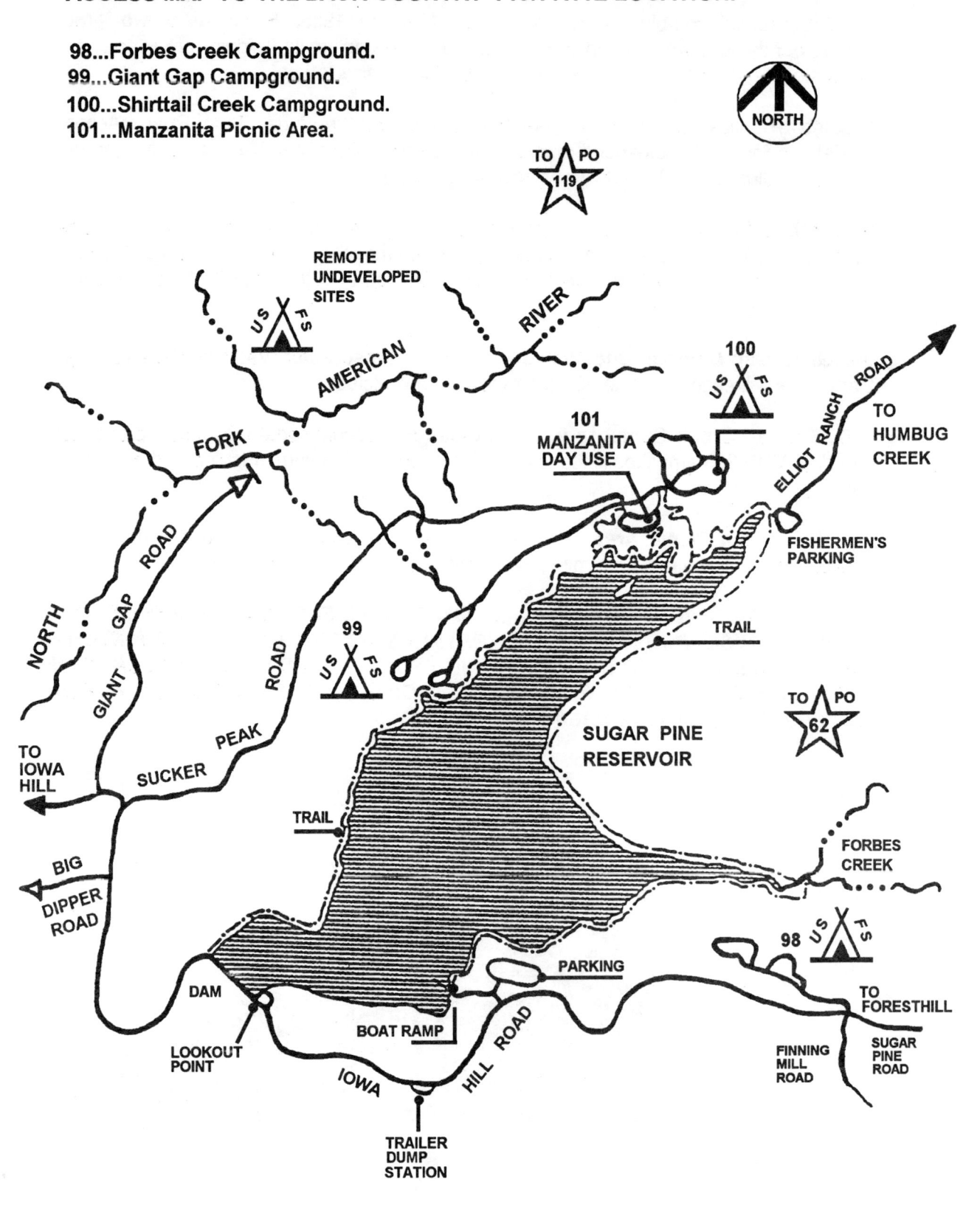

98...Forbes Creek Campground.
99...Giant Gap Campground.
100...Shirttail Creek Campground.
101...Manzanita Picnic Area.

FREE-USE RECREATIONAL GOLD MINING SITES-OPEN TO THE PUBLIC.

Every Forestry and BLM designated camp site are areas withdrawn from mineral entry. These sites can be used as FREE-USE RECREATIONAL GOLD MINING SITES, but cannot be claim staked.

Gold panning, the use of a Metal Detector, sniping for gold nuggets and the use of the rocker-sluice are activities which are permitted by the Forestry Service and the BLM within those Mineral Withdrawal camp site areas. Equipment is limited to a gold pan, associated sluices and small hand held tools. Operations are restricted to only those areas that are at water-level, or underwater. Motorized dirt moving equipment is prohibited.

Inquire with the area Forest Ranger and the BLM Supervisor before beginning serious mining activity with a suction-vacuum dredge. Different areas have varied regulations governing the use of suction dredges.

High-banking, removal of vegetation, relocation of natural configuration of the boulder structure within the water-run is prohibited. (In some areas boulders have been placed into position for the improvement of the fish habitat).

From time to time the Forestry Service and the BLM change their policies by removing the existing Mineral Withdrawal Areas from Public Free-Use by the order of the U.S. Congress.

This type of Congressional activity, at one time or another, will perhaps make some Mineral Withdrawal Sites here-in listed become obsolete. The Author has no control over those changes in the future with campsite locations as Mineral Withdrawal Sites that are in force at the time of this writing.

With over 3,000 known gold sites on hundreds of streams, water-runs and creeks in California, and an equal amount of undiscovered gold sites in the back-country, this Author lists only a small amount of those numbers. To do any differently would make this book of impossible dimension in size.

The gold yet to be located and harvested in California is estimated to be about 85% remaining, both in the placer streams and in the hard rock countryside.

SIERRA COUNTY *IS CLASS C, and is open to dredging from the fourth Saturday in May thru October 15.* Use the Tahoe National Forestry Map, County and California Highway Maps for locations.

Hydraulic mines had been worked for many years leaving behind as evidence to its work in piles 'n heaps of tailing debris. Over fifty-percent of its mining efforts fell to the debris heaps because of inefficient mining methods. Check those tailings for overlooked gold 'n platinum nuggets.

Placer mining areas at Downieville, Poverty Hill, Poker Flat, Port Wine, Brandy City, Fir Cap Mountains, China Flat, Crazy Croft and Slug Canyon produced considerable hydraulic mines with extensive tailings. Adjacent water-runs continue to give up placer gold that has been placed into the riffles by thunderstorms raking the surrounding back-country gravel of its gold content.

Patches of Tertiary quartz-rich gravel are known to exist in scattered areas of hundreds of feet wide and several miles in length. The adjacent creeks receive its gold content each season from winter storms disturbing the countryside of its gold particles and distributing these into riffle holding basins.

There are considerable mining claims dating back to the 1800's scattered among patented and private property. To allow the average gold recreational miner to share in the opportunity of retrieving the gold occurrences sent into the surrounding creeks and tributaries, the Forestry Service has set aside several Mineral Withdrawal Areas along Highway # 49 (the Goiden Highway), in the North Yuba River corridor.

Within each of those established sites, the recreational gold miner is permitted Free-Use to dredge, sluice-rocker, pan and nugget snipe with a Metal Detector without infringing on private property and established mining claims. Within those areas, mining claiming is prohibited. A California dredging permit is required. Gold panning is a Free-Use activity and does not require a permit in the State of California.

The following pages describe and illustrate several Mineral Withdrawal Areas set aside by the Forestry Service and list regulations governing the use within those Free-Use Areas.

Oregon Creek, Indian Valley, Convict Flat, Ramshorn, Union Flat, Wild Plum-Haypress Creek are all managed by the Tahoe National Forest Service as Free-Use Recreational Gold Mining Withdrawal Areas.

FREE-USE MINERAL WITHDRAWAL SITES--OPEN TO THE PUBLIC.

The Downieville Ranger District of the Tahoe National Forest has set aside the following six areas as Free-Use Mineral Withdrawal Sites for casual mining activities. Rules have been established to protect the many resource values along the North Fork of the Yuba River and to minimize conflict with other use-activities.

Each of the following hand drawn maps by the Author, indicated the type of mining activity allowed for a particular area along the North Fork of the Yuba River. **Oregon Creek, Indian Valley Convict Flat, Ramshorn, Union Flat and Wild Plum-Haypress Creek.**

Dredging permit is required for each dredge operator and can be obtained at the Region # 2 Headquarters, California Department of Fish and Game, 1701 Nimbus Road, Rancho Cordova, 95670.

Gold panning, sluicing and nugget sniping are Free-Use activities and do not require a permit in California, except, in special interest and sensitive areas as indicated elsewhere in California. Best to check with the local Ranger District in the area of interest.

For further information concerning the following Free-Use Mineral Withdrawal Areas, contact; North Yuba Ranger Station, 15924 Highway # 49, Camptonville, CA., 95922.   916-280-3231.

Casual Mining Rules include protection for the shoreline vegetation, the riparian area, the shoreline itself and highway fill slopes as well as preservation of the visual quality and visitor safety, requires that;
1...All mining activities will be restricted to the wetted perimeter of the existing natural flowing, stream channel.
2...There shall be no stream banking excavation.
3...There shall be no disturbance of rooted or embedded woody plants, trees, shrubs etc.
4...All work shall be performed by hand tools only.
5...Materials too large to be moved by hand are to be left as is.
6...Diversion of the flowing stream shall be only that necessary to direct water into the sluice box or to the extent necessary to operate a dredge and shall be removed prior to leaving the site.
7...The stream shall be returned to a condition similar to its original configuration prior to leaving the site.

8... Sluice boxes shall not be wider than two feet in the riffle area and not to exceed 50% of the wetted perimeter in length.

9... There shall be no hydraulicing (jet or nozzle), outside of the wetted perimeter.

10... Extreme care shall be taken to assure that no petroleum products or other deleterious material is allowed to fall, be wasted into, or otherwise deposited so as to enter surface waters.

11... You may conduct mining operations for 14 days per area, with a maximum use of 28 days per year if more than one area is used.

12... In the fall of 1989, boulders were placed in the Indian Valley Recreation Area to improve the fish habitat. To protect this investment and wildlife values, large rocks and boulders may not be moved within the North Yuba Fish Habitat Improvement Project Area.

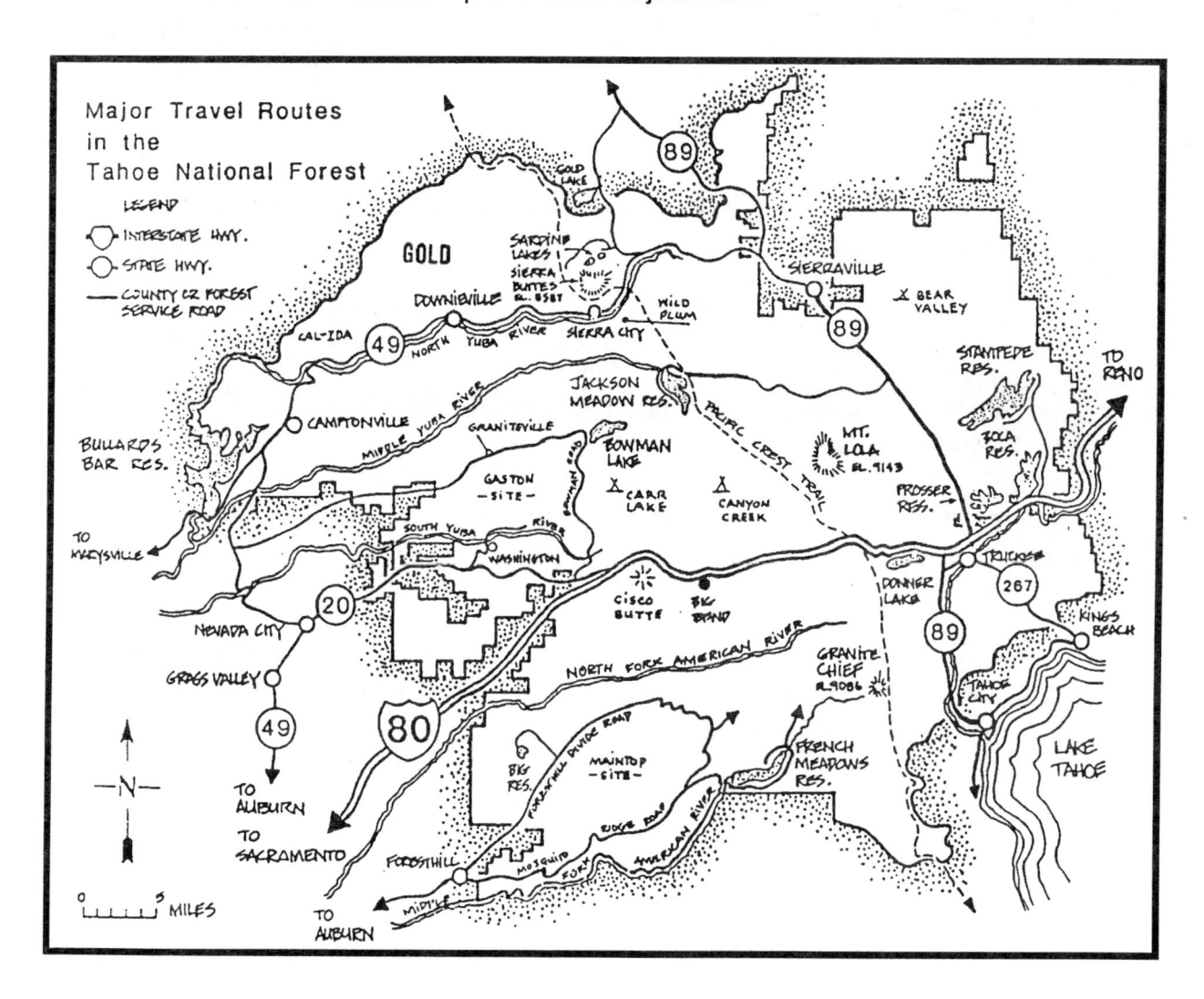

OREGON CREEK, T18N R8E Sec. 28. Use maps for USGS Topographic *Camptonville,* Tahoe National Forest Map, California Highway Map and Sierra County Map for location. Travel west and northwest on Highway # 49 (The Golden Highway), for about 17-miles to the river junction of the Middle Fork of the Yuba River and Oregon Creek.

At this **FREE-USE MINERAL WITHDRAWAL SITE, RECREATIONAL GOLD MINING** is allowed at 100 feet south of the Highway # 49 bridge that spans the Middle Fork Yuba River. The junction of the water-runs create a natural settling place for gold particles to accumulate in impressive numbers.

Private land surrounds this FREE-USE AREA. Restrict the gold mining activity to within the confines of the Oregon Creek Free-Use Area. Check with the North Yuba Ranger Station in Camptonville, California, for additional information.

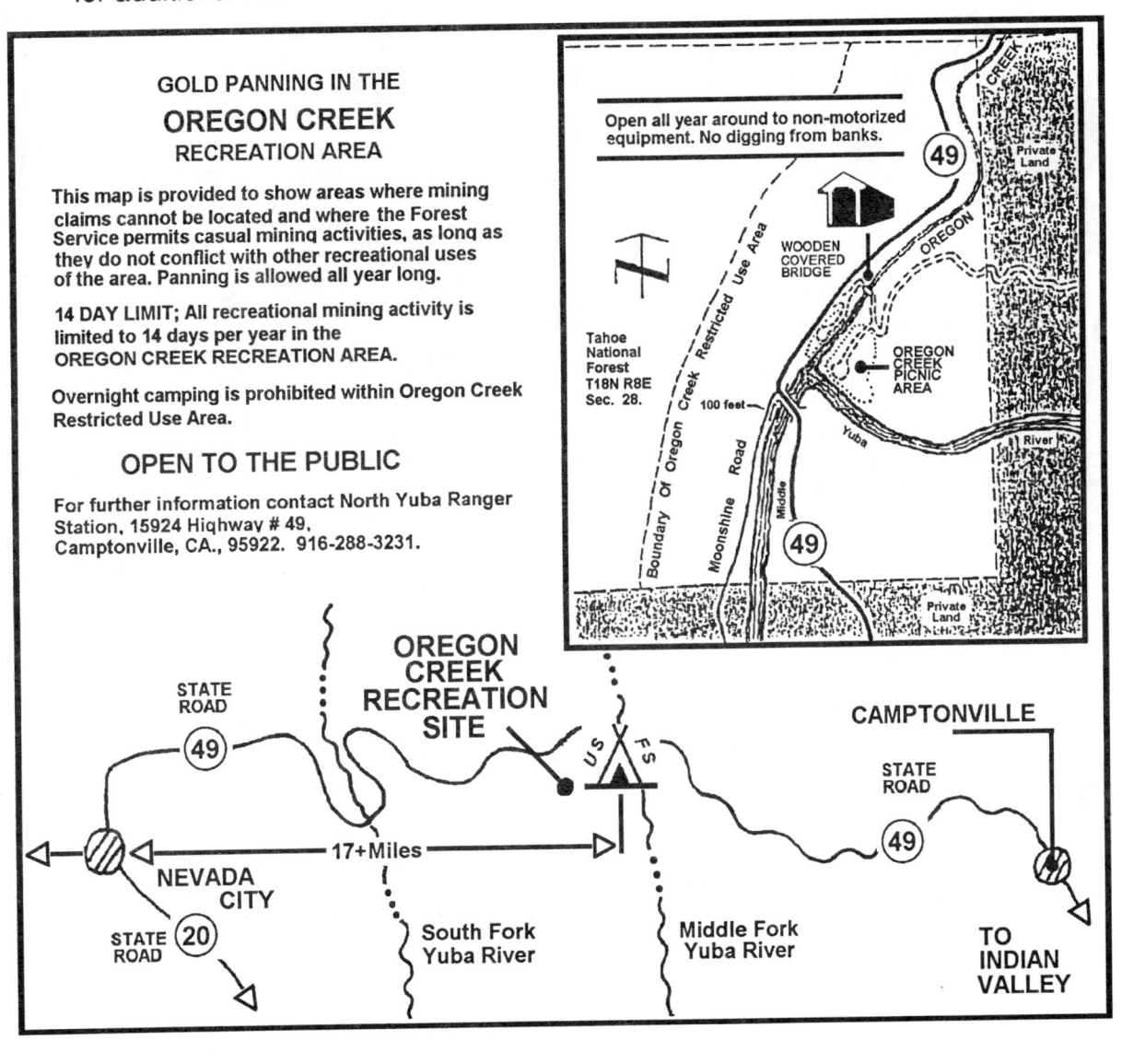

GOLD PANNING IN THE OREGON CREEK RECREATION AREA

This map is provided to show areas where mining claims cannot be located and where the Forest Service permits casual mining activities, as long as they do not conflict with other recreational uses of the area. Panning is allowed all year long.

14 DAY LIMIT; All recreational mining activity is limited to 14 days per year in the OREGON CREEK RECREATION AREA.

Overnight camping is prohibited within Oregon Creek Restricted Use Area.

OPEN TO THE PUBLIC

For further information contact North Yuba Ranger Station, 15924 Highway # 49, Camptonville, CA., 95922. 916-288-3231.

50

INDIAN VALLEY T19N R9E. Use maps for Topographic *Goodyears Bar,* located along Highway # 49 east of Camptonville for about ten-miles. Good camping sites with easy access for campers.

Notable for gold occurrences on the North Fork of the Yuba River, especially after the snowmelt run-off and seasonal rain fall. Back-country is famous for those deeply hidden ancient Tertiary graveled riffles with rich High-grade pocket basins held in the Quaternary river channels.

Check the inside of the bends in the river for the gold particles, and where the channel straightens out....test the middle course of the river for gold occurrences. For additional information, contact; North Yuba Ranger Station, 15924 Highway # 49, Camptonville, CA., 95922. 916-288-3231.

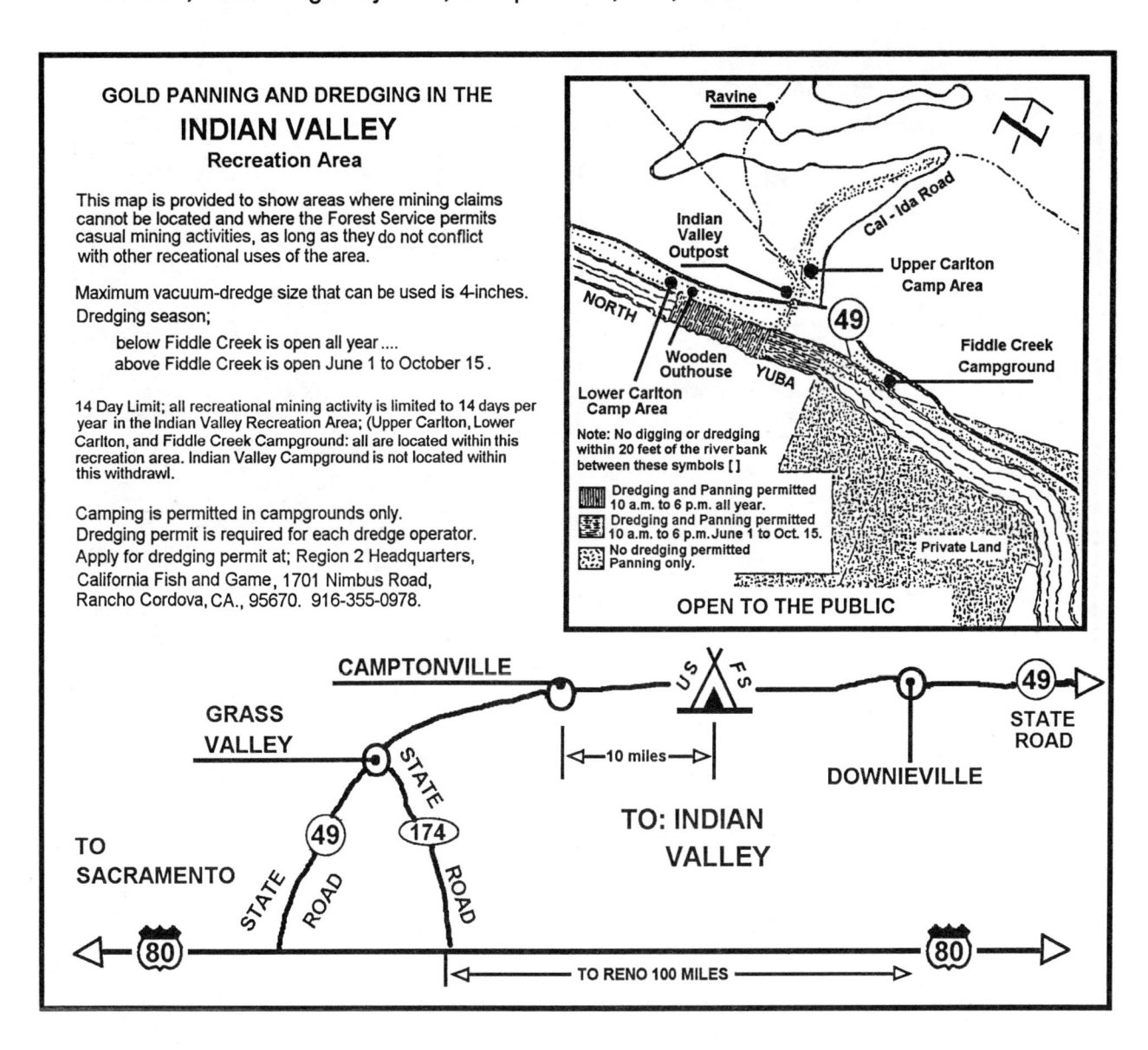

GOLD PANNING AND DREDGING IN THE
INDIAN VALLEY
Recreation Area

This map is provided to show areas where mining claims cannot be located and where the Forest Service permits casual mining activities, as long as they do not conflict with other receational uses of the area.

Maximum vacuum-dredge size that can be used is 4-inches.
Dredging season;

    below Fiddle Creek is open all year....
    above Fiddle Creek is open June 1 to October 15.

14 Day Limit; all recreational mining activity is limited to 14 days per year in the Indian Valley Recreation Area; (Upper Carlton, Lower Carlton, and Fiddle Creek Campground: all are located within this recreation area. Indian Valley Campground is not located within this withdrawl.

Camping is permitted in campgrounds only.
Dredging permit is required for each dredge operator.
Apply for dredging permit at; Region 2 Headquarters, California Fish and Game, 1701 Nimbus Road, Rancho Cordova, CA., 95670. 916-355-0978.

Note: No digging or dredging within 20 feet of the river bank between these symbols [ ]

- Dredging and Panning permitted 10 a.m. to 6 p.m. all year.
- Dredging and Panning permitted 10 a.m. to 6 p.m. June 1 to Oct. 15.
- No dredging permitted Panning only.

OPEN TO THE PUBLIC

TO: INDIAN VALLEY

TO RENO 100 MILES

51

CONVICT FLAT T19N R9E Sec. 10. Use maps for USGS Topographic *Goodyears Bar,* Tahoe National Forest Map, California and Sierra County Map. This is a FREE-USE MINERAL WITHDRAWAL SITE located one-mile east of the Rock Rest Campground on Highway # 49. The North Yuba River plays host to the Golden Highway # 49 for over fifty miles with recreational gold mining sites. Area produces favorable sized gold nuggets. This is a good area to nugget snip and to work a small back-packing suction dredge into isolated creek sites in the back-country. Watch out for those claim markers while in the back-country. There is a prominent large curve in the North Yuba River at this site, accentuated with exposed bedrock of cracks, fissures and catch-alls in which passing gold particles settle in. The bedrock is vacant of thick unmanageable overburden except for a few alternating sand bars and Tertiary gravel heaps. This makes for easier working conditions to obtain the gold that is deposited by heavy rains into the ancient channeled water-run.

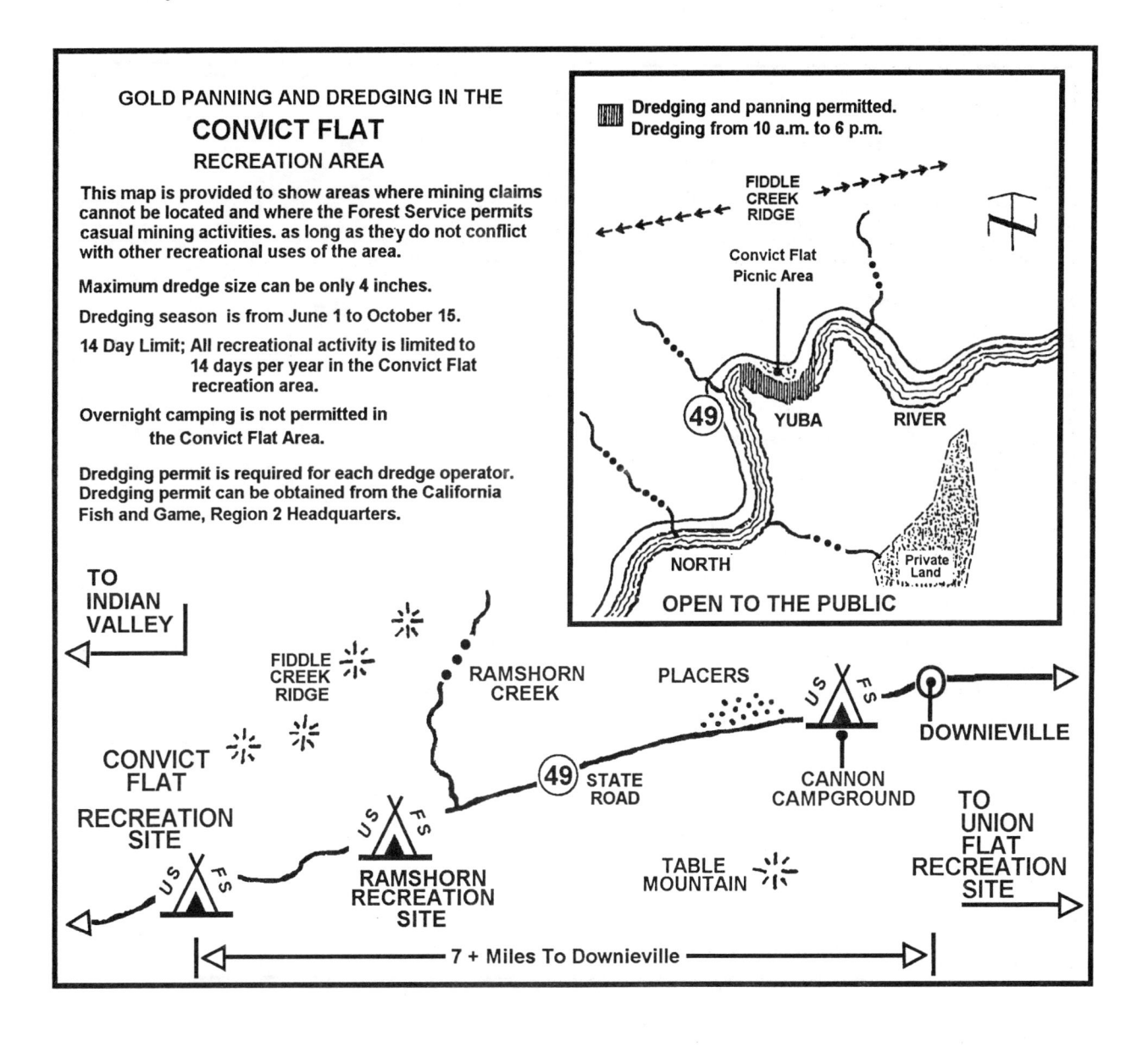

GOLD PANNING AND DREDGING IN THE
CONVICT FLAT
RECREATION AREA

This map is provided to show areas where mining claims cannot be located and where the Forest Service permits casual mining activities. as long as they do not conflict with other recreational uses of the area.

Maximum dredge size can be only 4 inches.

Dredging season is from June 1 to October 15.

14 Day Limit; All recreational activity is limited to 14 days per year in the Convict Flat recreation area.

Overnight camping is not permitted in the Convict Flat Area.

Dredging permit is required for each dredge operator. Dredging permit can be obtained from the California Fish and Game, Region 2 Headquarters.

Dredging and panning permitted. Dredging from 10 a.m. to 6 p.m.

FIDDLE CREEK RIDGE

Convict Flat Picnic Area

49 — YUBA RIVER

NORTH — Private Land

OPEN TO THE PUBLIC

TO INDIAN VALLEY

FIDDLE CREEK RIDGE

RAMSHORN CREEK

PLACERS

DOWNIEVILLE

CONVICT FLAT RECREATION SITE

49 STATE ROAD

CANNON CAMPGROUND

RAMSHORN RECREATION SITE

TABLE MOUNTAIN

TO UNION FLAT RECREATION SITE

|← 7 + Miles To Downieville →|

RAMSHORN T19N R9E Sec. 1. Use maps for USGS Topographic *Goodyears Bar,* Tahoe National Forest Map, California Highway Map and Sierra County Map. This FREE-USE MINERAL WITHDRAWAL SITE located two-miles plus, east of Convict Flat Recreational Gold Mining Site on Highway # 49. The Ramshorn Recreation Area was named after the Ramshorn Creek which dumps into the North Yuba River just opposite the island that sets in the Yuba River proper. This area can be located on the Tahoe National Forestry Map at T19N R9E/R10E Sec. 1.

Silvery looking nuggets that are found in the water-run are in reality gold nuggets coated with mercury that the old-timers had used in their inefficient mining methods. Excessive amounts of mercury had splattered into the gravel of the water-run from over-pour into the sluice's riffles by the old-miners in hoping that "more was better". The mercury overfilled riffles caused the lighter gold to float 'n glide on top of the mercury to flow right back into the creek proper and become amalgamated substances. **Do subject research before processing.**

GOLD PANNING AND DREDGING IN THE

RAMSHORN
RECREATION AREA

This map is provided to show areas where mining claims cannot be located and where the Forest Service permits casual mining activities, as long as they do not conflict with other recreational uses of the area.

Maximum dredge size that can be used is four inches.

Dredging season is from June 1 to October 15.

14 Day Limit; All recreational mining activity is limited to 14 days per year in the Ramshorn Recreation Area.

Camping is permitted in campgrounds only.

Dredging permit is required for each dredge operator. Dredging permit may be obtained from the California Fish and Game Management of Region 2.

THIS FREE-USE SITE IS OPEN TO THE PUBLIC

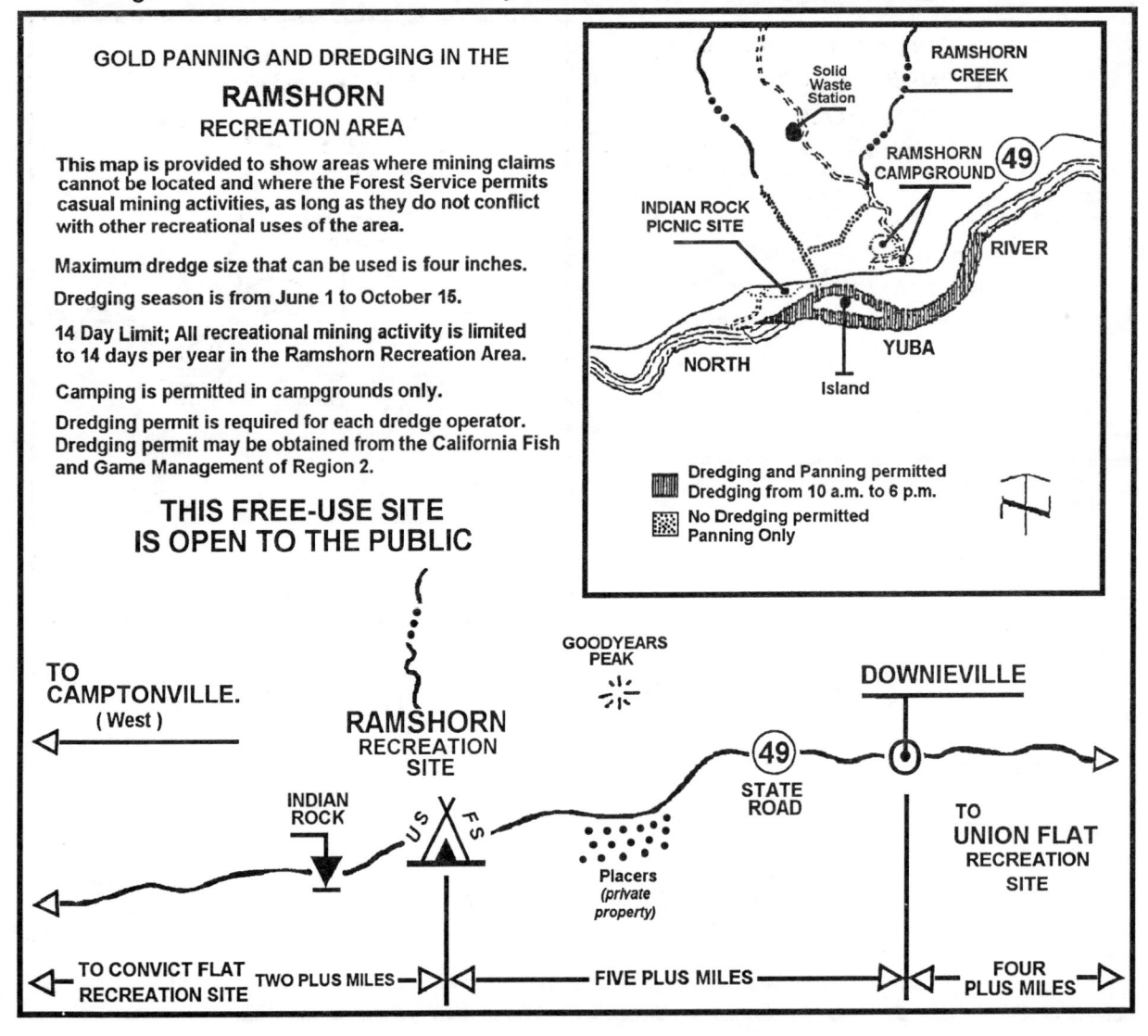

- Dredging and Panning permitted Dredging from 10 a.m. to 6 p.m.
- No Dredging permitted Panning Only

TO CAMPTONVILLE (West) ← | RAMSHORN RECREATION SITE | GOODYEARS PEAK | 49 STATE ROAD | DOWNIEVILLE | → TO UNION FLAT RECREATION SITE

INDIAN ROCK | USFS | Placers (private property)

← TO CONVICT FLAT RECREATION SITE — TWO PLUS MILES — FIVE PLUS MILES — FOUR PLUS MILES →

UNION FLAT T20N R11E Sec. 28 and 33. Use maps for USGS Topographic *Sierra City;* Tahoe National Forest Map, California Highway and Sierra County Map. This FREE-USE MINERAL WITHDRAWAL SITE is located four-miles plus, east of Downieville right on the North Yuba River proper and on the Golden Highway no. 49. Easy access to camping sites and to recreational gold mining localities. Union Flat Campground is located opposite the Forestry Road no. 93 and in between Highway no. 49 and the North Yuba River proper.

Gold dredging begins at the west end of Union Flat Campground where the tall leaning pine tree sets brooding. Dredging ends at the eastern end of the campground near the retaining wall and ends at the Ox-bow near the private property line. In between the campground and the private property area, dredging is permitted at China Flat Site.

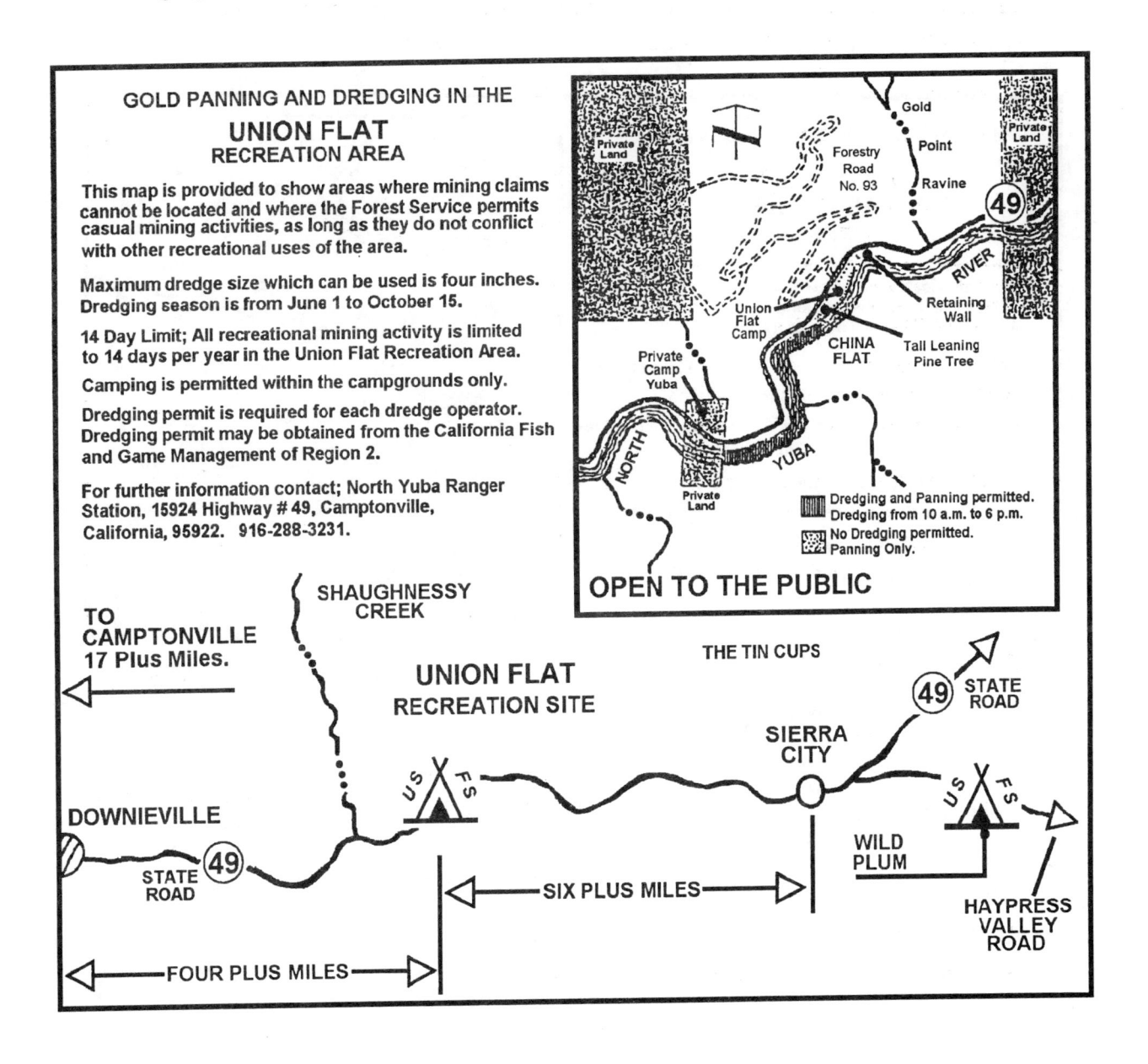

GOLD PANNING AND DREDGING IN THE UNION FLAT RECREATION AREA

This map is provided to show areas where mining claims cannot be located and where the Forest Service permits casual mining activities, as long as they do not conflict with other recreational uses of the area.

Maximum dredge size which can be used is four inches. Dredging season is from June 1 to October 15.

14 Day Limit; All recreational mining activity is limited to 14 days per year in the Union Flat Recreation Area.

Camping is permitted within the campgrounds only.

Dredging permit is required for each dredge operator. Dredging permit may be obtained from the California Fish and Game Management of Region 2.

For further information contact; North Yuba Ranger Station, 15924 Highway # 49, Camptonville, California, 95922. 916-288-3231.

OPEN TO THE PUBLIC

Dredging and Panning permitted. Dredging from 10 a.m. to 6 p.m.
No Dredging permitted. Panning Only.

54

WILD PLUM--HAYPRESS CREEK T20N R12E Sec. 26 and 27.
Use maps USGS Topographic *Haypress Valley;* Tahoe National Forest Map, California Highway Map and Sierra County Map.

Situated about ten-miles east of the town of Downieville on the Golden Highway # 49, sets an old mining community, Sierra City, from here travel about one half-mile east to the turn-off onto the Haypress Valley Road, travel one-mile south in this direction to where the Wild Plum--Haypress Recreation Site is located.

Haypress Creek is a long tributary of the Yuba River with a history of placer mining. Seasonal gold deposits from the remnants of the Tertiary gravels find their way into the Haypress Creek. It is believed that beneath the present overburden, there exists in staggered locations those same nugget deposit formations that once laid openly on the surface, such as where the ancient "Tin-Cup" deposits now exist just west of the community of Sierra City.

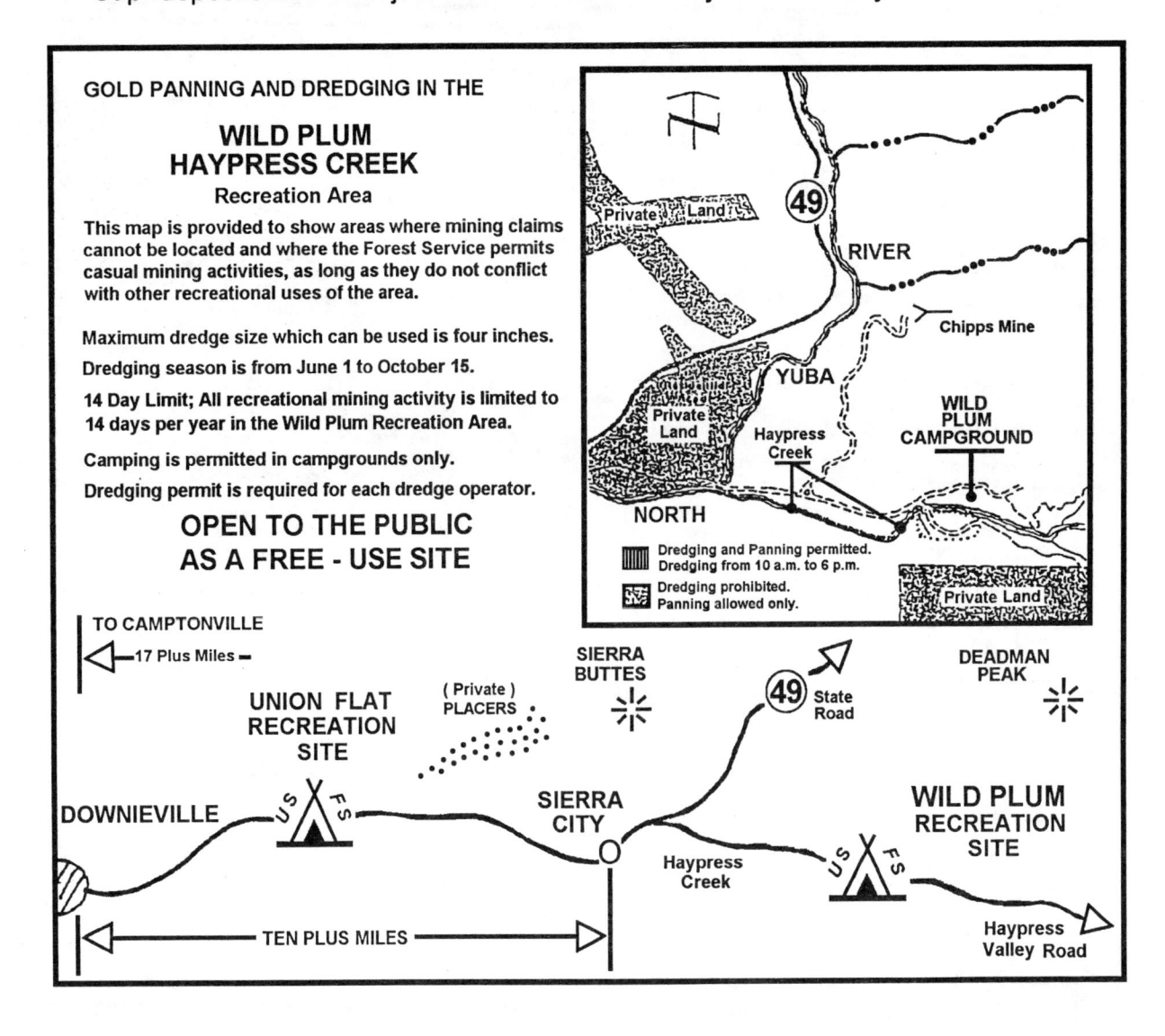

GOLD PANNING AND DREDGING IN THE
WILD PLUM HAYPRESS CREEK
Recreation Area

This map is provided to show areas where mining claims cannot be located and where the Forest Service permits casual mining activities, as long as they do not conflict with other recreational uses of the area.

Maximum dredge size which can be used is four inches.

Dredging season is from June 1 to October 15.

14 Day Limit; All recreational mining activity is limited to 14 days per year in the Wild Plum Recreation Area.

Camping is permitted in campgrounds only.

Dredging permit is required for each dredge operator.

OPEN TO THE PUBLIC AS A FREE - USE SITE

Dredging and Panning permitted. Dredging from 10 a.m. to 6 p.m.

Dredging prohibited. Panning allowed only.

55

TRINITY COUNTY; CLASS E. Open to dredging from July 1 through September 30, except where noted. Use Shasta-Trinity National Forest Map, County and California Highway Map for location into the back-country. Canyon Creek has extensively been worked for about 12 miles of its length. This gives the metal detectorist an opportunity to scour the debris heaps for overlooked gold particles. Be alert to the land status before venturing into suspicious tailings. There are many old tailing piles that no longer are covered by mining claims and can be nugget sniped.....the detectorist must know where these are by doing research work.

Straddling the Trinity-Shasta County line at location; T33N R8W, past history of lode mining became the most important district in the Klamath Mountains. The area tributaries have yielded large amounts of placer gold that had been supplemented by consistent rain storms raking the huge surface ore bodied.

At the foot of the old mines the float from its debris dumps give credence to what has been mined. Many instances prove with a metal detector that the old miners had overlooked gold particles of varying sizes that had been sluiced onto the tailing piles. Check the ore dump at the old mines, especially where the mineralized shear zones are in and along the contacts of lamprophyre dikes that cut and intercept serpentine formations. The back-country has been known to give up evidence of quartz stringers holding free gold associated with abundant sulfides. The ravaging winter storms rake the quartz stringers, freeing these into the riffle holding basins in the adjacent tributaries. High-grade pockets were discovered in the past and continue to be a possibility of existence, deserving serious exploratory consideration and attention. The major water-runs and their feeder creeks continue to give up placer gold occurrences from their intimidating back-country and lure today's dredgers to work the aggregate 'n gravel.

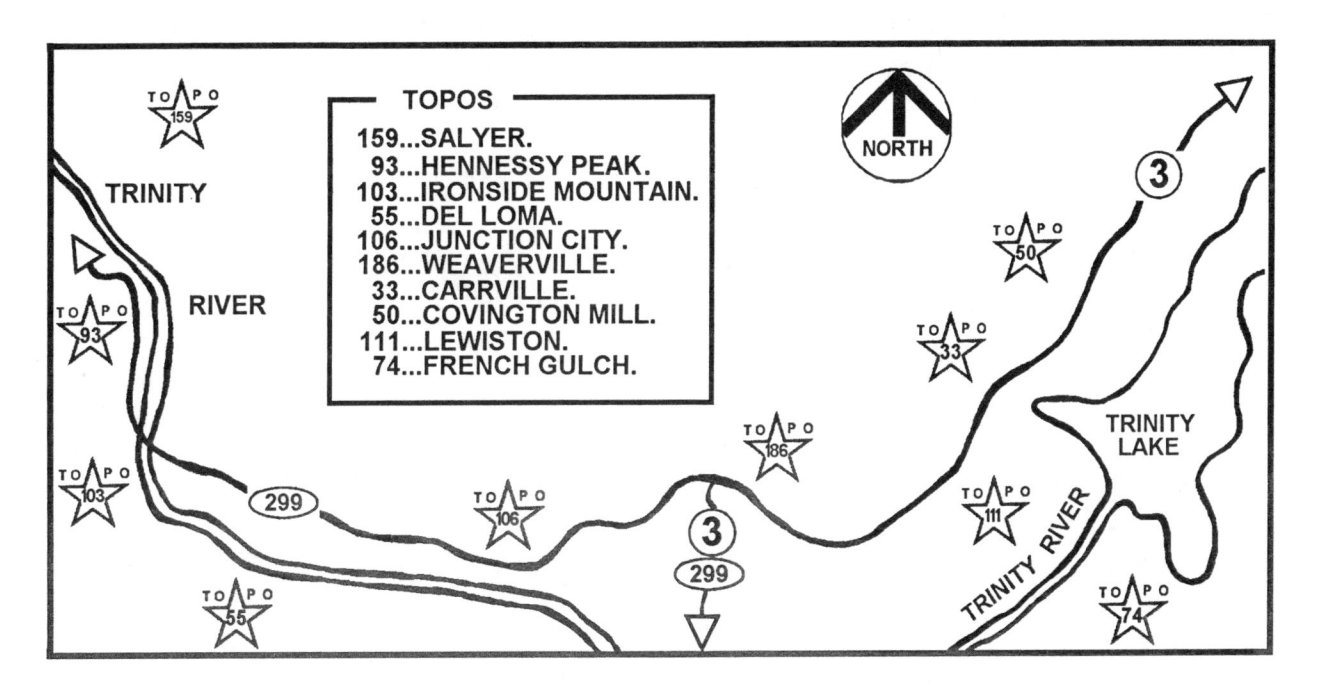

TOPOS
159...SALYER.
93...HENNESSY PEAK.
103...IRONSIDE MOUNTAIN.
55...DEL LOMA.
106...JUNCTION CITY.
186...WEAVERVILLE.
33...CARRVILLE.
50...COVINGTON MILL.
111...LEWISTON.
74...FRENCH GULCH.

JUNCTION CITY CAMPGROUND.
USGS TOPO; *JUNCTION CITY T5N R10W AND R11W.*
KLAMATH RIVER CAMP SITE. USGS TOPO; *COPCO T48N R4E.*

For getting around the back-country water-runs use the California Highway Map, Shasta-Trinity National Forest and the Klamath National Forest Map.

JUNCTION CITY CAMPGROUND has low key gold panning opportunities. Steep incline to river access with plenty of vegetation in the river proper due to damming the river. Beneath the vegetation sets pockets, basins and dried up riffles that are holding gold occurrences that once were deposited by the flow of the river into those natural receptacles. A good gold metal detector would be the equipment to use for sniping for those pockets of gold nuggets.

The landscape holds "birdeye porphyry" dikes in gold quartz veins and ore shoots where the seasonal rains percolate the land and deposit those gold particles into the many tributaries that flow through the BLM lands.

With the Forestry Wilderness Areas off-limits to any type of mineral extraction, *and these surround the BLM lands*, the build up of gold occurrences in the many water-runs continue because of the lack of mining activity. The gold is moved and sent its way by run-off into the many BLM water-runs and their tributaries.

Junction City Campground will serve as a base camp for the gold searcher as they enter into the BLM back-country for spot checking and performing test holes in the old channels, terrace gravels and existing water-run gravel aggregate. *Be alert to existing mining claims in the area.*

KLAMATH RIVER CAMP SITE is a BLM managed primitive camp located about 6.5 miles north of Copco Lake on Topsy Grade Road. Primarily used by fishermen, but the gold searcher can use it as a base camp for exploring the back-country and those of near-by Oregon's water-runs. Exit U.S. Highway no. 5 east of Hornbrook and follow the road alongside of the reservoir to the area of interest.

Klamath River is joined by Cottonwood Creek at Hornbrook, which was made famous by its very rich 'n shallow gold placer deposit mining of the past. This is indicative of the kind of land area where shallow terraces, channel gravels and ancient dried up rivers lay beneath the grasslands, oak and pine topography. Big boulders occupy this section of the Klamath River.

WHERE THE OLD MINERS ONCE MINED FOR PLACER GOLD IS STILL A GOOD AREA TO PROSPECT.

Junction City campground is a good jump-off point into the BLM lands that host ancient terraces & channels holding gold nuggets.

BLM lands are numerous in this area and are surrounded by Forestry Wilderness Areas that prohibits mineral extraction from its water-runs.

About 12 miles west from Junction City Campground on Highway 299 sets a Free-Use Site, BIG FLAT RECREATIONAL GOLD MINING SITE. See following pages.

For information about these areas check with BLM Redding Office at; 916-224-2100.

For information about FREE camping areas on BLM lands in the West, call; 1-800-47-SUNNY.

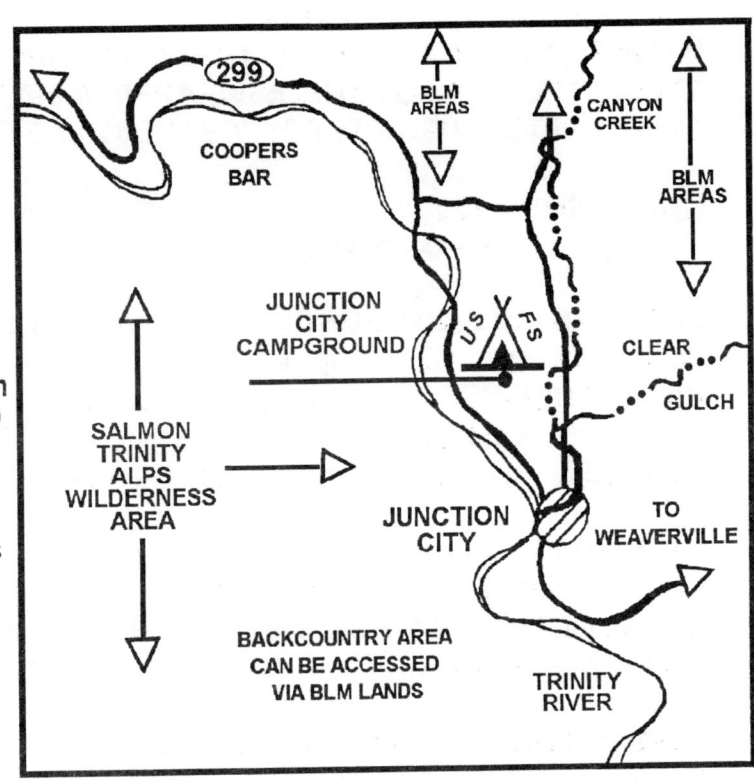

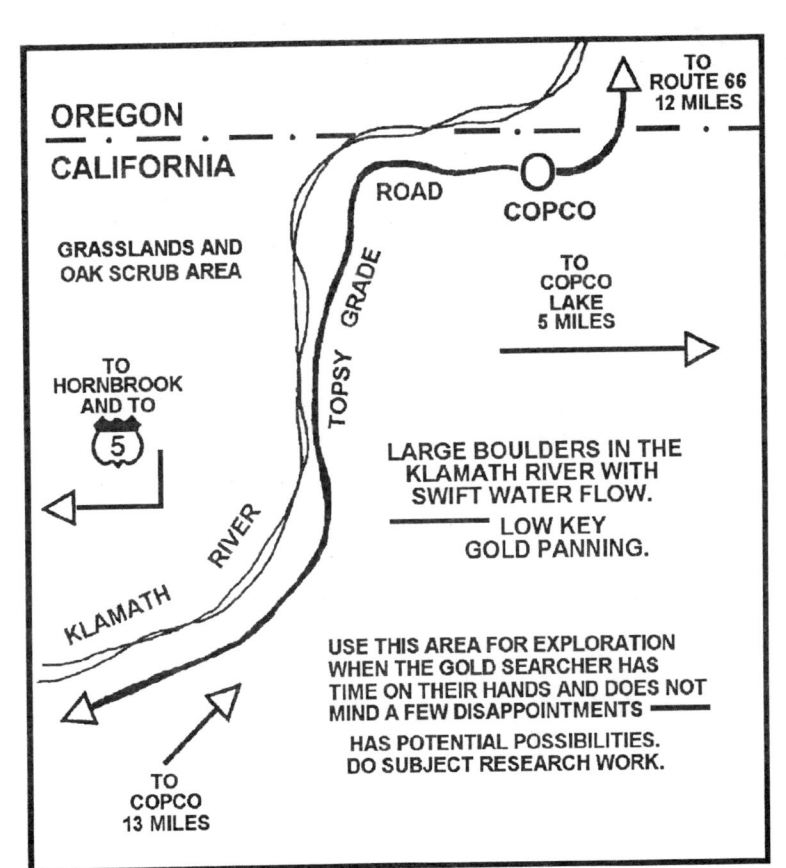

Klamath River enters that portion of the Klamath Mountain area at Hornbrook, where it flows southwest and then west for about 50 miles, crossing through a number of mining districts gathering and carrying gold occurrences as it travels.

Present placer mining activity has been carried on in the river channel where it had cut thrue ancient terraces, stream-beds, graveled gulches and benches.

The present Klamath River has cut thrue those ancient deposit debris that extend for miles. The younger water-runs and their tributaries have flowed across and thrue those aggregate deposits and created new deeper channels sending the gold particles into newer riffle basins.

DOUGLAS CITY CAMPGROUND.
USGS TOPO; *WEAVERVILLE T32N R10W Sec. 10.*
STEINER FLAT. USGS TOPO; *WEAVERVILLE T33N R10W Sec. 35.*
STEELBRIDGE. USGS TOPO; *WEAVERVILLE T33N R9W Sec 28/32.*

Use California Highway Map and Shasta-Trinity National Forestry Map for Getting around the countryside.

DOUGLAS CITY CAMPGROUND. Low key gold panning. Exit one-half mile West of Douglas City along the Trinity River on Highway 299 to Steiner Flat Road. FREE-USE fully developed campground. Mechanized equipment prohibited. Hand-held tools, rocker-sluice, metal detector and gold pan is permissible. BLM public lands. 14 day use. May 1--Nov. 1.

STEINER FLAT . Low key gold panning. Primitive camping and river access. No use fee. BLM public lands. 1.7 miles NW of Douglas City Campground located on Steiner Flat Road along the Trinity River. All year use season. Hand-held tools only. Rocker-sluice, gold pan and metal detector is permissible.

STEELBRIDGE CAMPGROUND. Fair gold panning. Two locations on the Trinity River adjacent to one another. Located 3 miles east of Douglas City on Highway 299, then travel 2 miles north on the Steelbrdge Road. No use fee. 14 day stay limit. Primitive camping. Use season all year. Gold pan, rocker-sluice, hand-held tools and metal detector is the recommended equipment for use on the BLM public lands in this area.

The major tributaries of the Trinity River are Stewart's Fork, East Fork, Coffee Creek, Indian Creek, New River, Hayfork Creek, Big French, Manzanita and Sailor Bar Creek.

Gold is found not only in the gravel in the present stream channels, but also in the older terrace and bench deposit contingent to the channels. One of the largest hydraulic placer mines that was operated in California years ago sets a few miles west of Weaverville, as a reminder of its heaping tailings. Use a metal detector to check those tailings as the old miners left behind 55% of their gold particles due to inefficient mining methods. Many gold nuggets remain in the tailing piles which were seen by the old miners as being swept away by their fast moving sluices but could not be retrieved.

For further information, write to; BLM Redding Resource Area, 355 Hemsted Drive, Redding, CA, 96002-0910. 916-224-2100.

Large amounts of GOLD have been taken from the tributaries of the area in the last Century.

After each rain storm, part-time gold snipers become active in sweeping the water-runs for the GOLD particles that have been washed into the gravel.

Watch for platinum and its five member family in the creeks and on the old mine tailings.

Search the magnetite sand for the gold fines. Wash the concentrates two and three times over until the color shows.

Search the tailing piles for gold and platinum nuggets with the metal detector.

Be alert to mining claims.

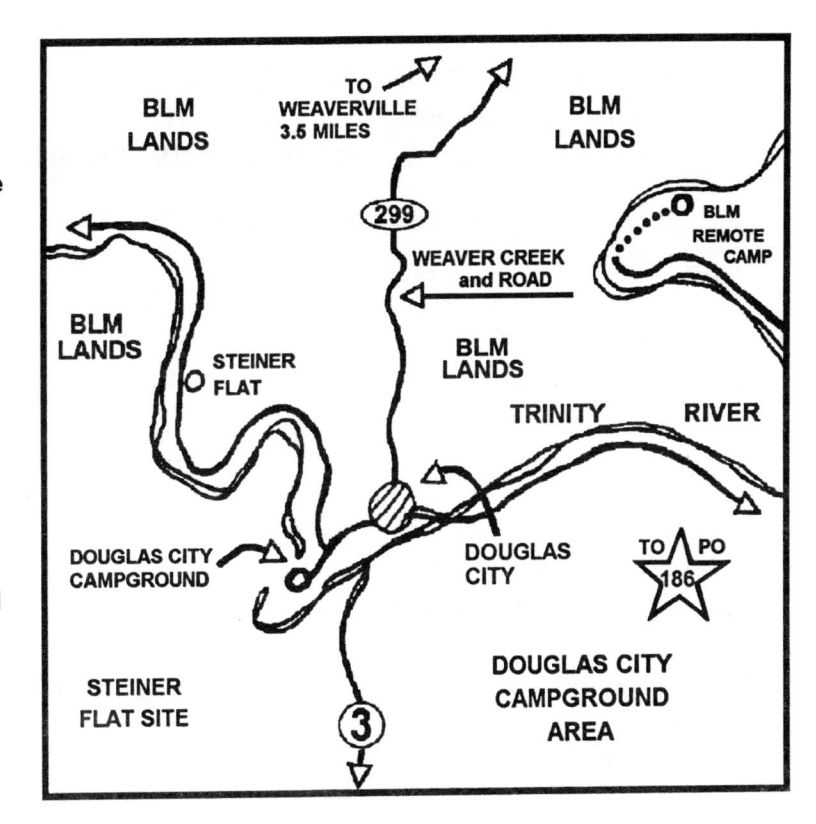

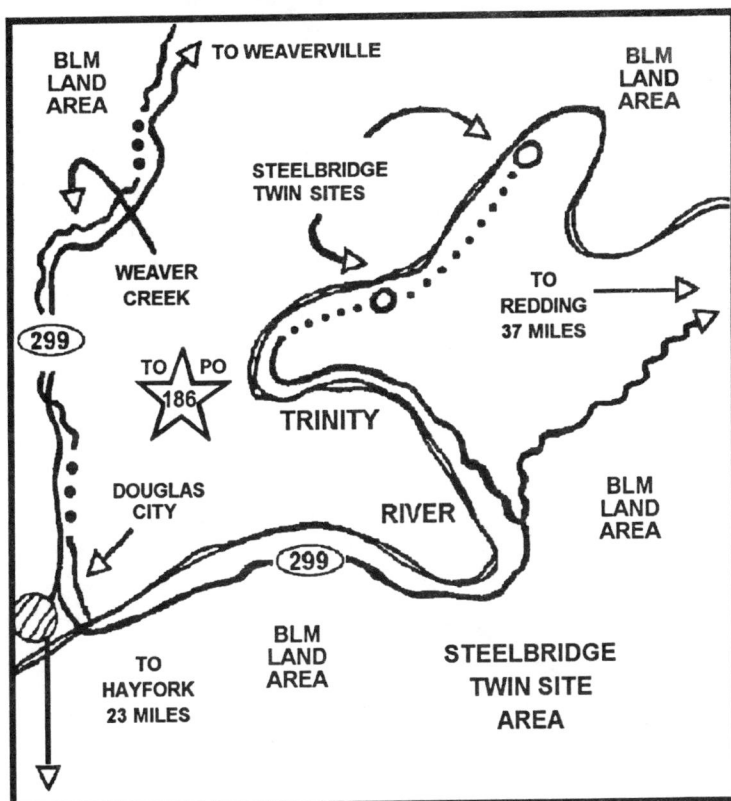

DOUGLAS CITY CAMPGROUND AREA

STEINER FLAT SITE

The BLM lands of this area has a history of rich placer gold mining.

NW of Weaverville for about 15 miles is located Canyon Creek, which had a furious mining history in the past. Hydraulic mining and its tailing piles extend for over 12 miles.

The back-country offers many opportunities for setting down test holes in the many untouched bench gravels that host the ancient gold nuggets.

Metal detecting and gold sniping any old mine ore dump debris and placer mine tailing piles will bring pleasant results.

Use a Shasta-Trinity National Forest Map for locating the areas of interest.

STEELBRIDGE TWIN SITE AREA

60

BIG FLAT - BIG BAR CREEK. MINERAL WITHDRAWAL SITE.
USGS TOPO; *HAYFORK BALLY AND BIG BAR.*
T4N R12W Sec. 3. Shasta - Trinity National Forest Map.

Big Flat Mineral Withdrawal Site extends for about 300 yards along the Trinity River where the Big Bar River dumps into it. Use within the Big Flat Area is limited to seven recreational dredging miners at one time and is available for up to 14 days Free-Use for each individual for the seasonal year, and this is based on first-come first-served. **OPEN TO THE PUBLIC.**

To reach this Mineral Withdrawal Site from Weaverville, travel west on State Highway no. 299 for about 21 plus-miles. The Recreation Use Area will be located in Big Flat between the store (*TRINITY RIVER INN*) and a private cable car that crosses the river at *Big Bar River (Creek.)* This is a Mineral Withdrawal Site where filing a mineral claim is prohibited.

To request the Free-Use Permit and the informative brochure entitled, "Big Flat Free-Use Permit", write to; Big Bar Ranger Office, Star Route No. 1, Box 10, Big Bar, CA., 96010.   916-623-6106.

Should the recreational miner decide to use dredging gear in this Free-Use Site, they will be required to obtain a dredging permit from the California Department Of Fish and Game, 601 Locust St., Redding, CA., 96001.   916-225-2300.

The area has a history of **PLACER GOLD** mining and is located right on one of those many extensions belonging to the, "Lost Blue Continents", of which the Yuba Gold Country is a family member. The campsites can be used for a base for jumping off into the back-country to explore for the many gold occurrence possibilities.

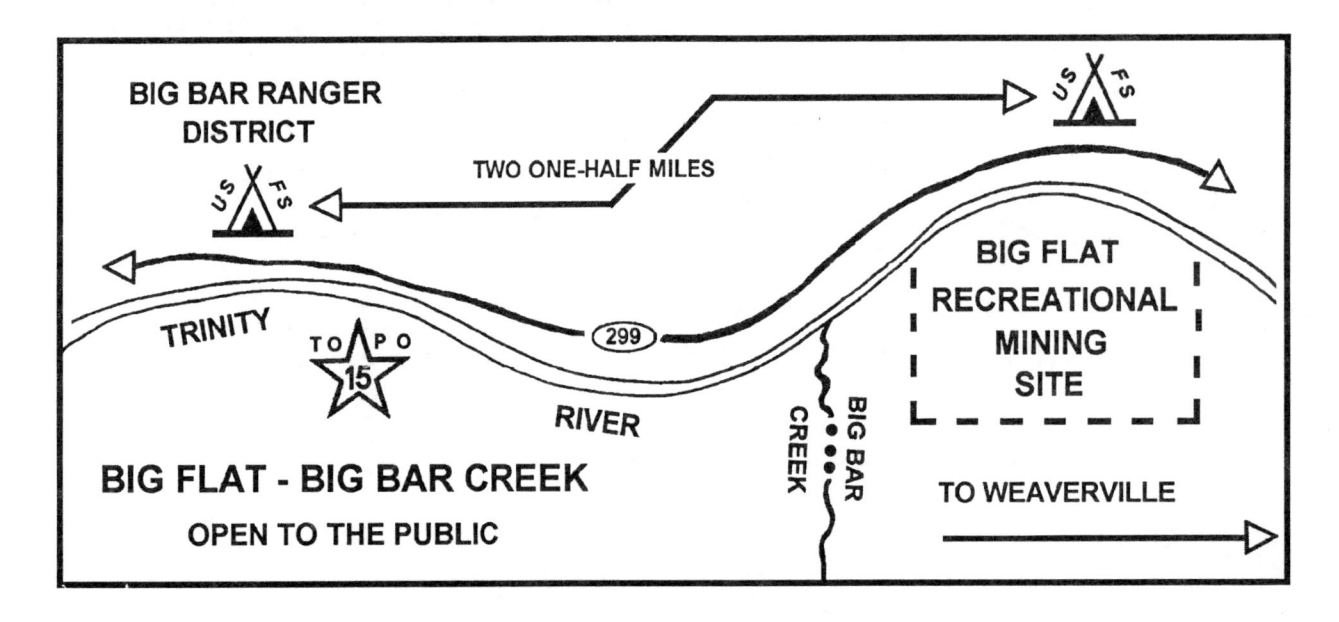

SOUTH YUBA WILDERNESS. OPEN TO THE PUBLIC AS A FREE-USE RECREATIONAL GOLD MINING CORRIDOR.
EDWARDS CROSSING; T17N R9E Sec. 17 TOPO; *WASHINGTON.*
MISSOURI BAR; T17N R10E Sec. 8. TOPO; *NORTH BLOOMFIELD.*

BLM managed area. A California dredging permit is required. Gold panning, sluicing, gold nugget sniping and the use of a rocker-sluice are a Free-Use activity in California where permitted.

The campsite at Edwards Crossing is situated on an ancient river-bed that was hydraulic mined leaving tailings showing as evidence of having been worked years ago. The old-timers with their inefficient mining methods overlooked as much as 55% of the gold that was lost to the tailing piles. Gold nugget sniping with a metal detector in this area is a worthwhile activity. Most of the trail from Edwards Crossing is about 700 feet above the river proper following the Canyon escarpment for about five miles to Missouri Bar, a primitive campsite that can be used as a base camp where the gold searcher can enter the back-country to search for gold particles.

Access from the trail down to the river at several points is a test of ingenuity and perseverance because of the rugged wilderness terrain. The South Yuba River flows over and through ancient placer gold territory as evidenced by the old North Bloomfield, North Canyon and Malakoff hydraulic diggings.

With seasonal thunderstorms, flooding and gravel movement redistributing much of the ancient gold along the river main, there most likely are remote areas where the river bedrock has never seen a dredge at work or explored with test holes. Most all of the crevices, fissures and cracks in the bedrock captures and holds gold nuggets that the dredge cannot move and requires a hand-held tweezers to extract them. Approximately one-third of the gold take comes from the fractured area of the bedrock.

To find a suitable dredging site with camping area requires considerable hiking into the wilderness along the South Yuba River flow. Much of the area hasn't been visited in over 100 years because of the primitive location, and where people don't frequent an area, the wildlife and vegetation flourishes. Take caution where one places their hands and feet, as this is rattle snake and poison oak-ivy country.

A good period to hit the river banks, is right after a heavy rain fall that created humungus surges of water flow. Percolation of the land loosens the Tertiary placer gold and sends this into the natural riffles of the river main where a dredge can go to work on the drop points. A dredging permit may be obtained from the California Fish and Game; 3211 "S" St., Sacramento, CA., 95816.

SOUTH YUBA RECREATIONAL GOLD MINING AREA.
USGS TOPO; *NORTH BLOOMFIELD.* **T17N R9E Sec. 13 thrue 16/22.**
Use California Highway Map, Tahoe National Forest Map, Thomas Street Guide for Nevada County.

The SOUTH YUBA RIVER RECREATIONAL GOLD MINING AREA is managed by the BLM for the Public recreational gold mining use. The filing of a mineral claim is prohibited in this Free-Use area. The Mineral Withdrawal Area is located some 12 twisting miles northeast of Nevada City by traveling Highway # 49 to Tyler Foote Crossing Road. Dredging and camping are limited on the South Yuba River to those areas that are posted and referred to within the rules of the land brochure furnished by the BLM. For information write to; BLM Folsom Resource Area, 63 Natoma St., Folsom, CA., 95630. 916-985-4474. South Yuba River receives its share of the gold from the tributaries of the surrounding area during rain run-off from rain storms that rake and loosen those ancient channel gravel deposits exposed on the surface. A west-held Tertiary channel extends to North Bloomfield with overplay of andesite gravel and branches out with the non-plus Blue Tent influence to the North Columbia placer gold-platinum deposits.

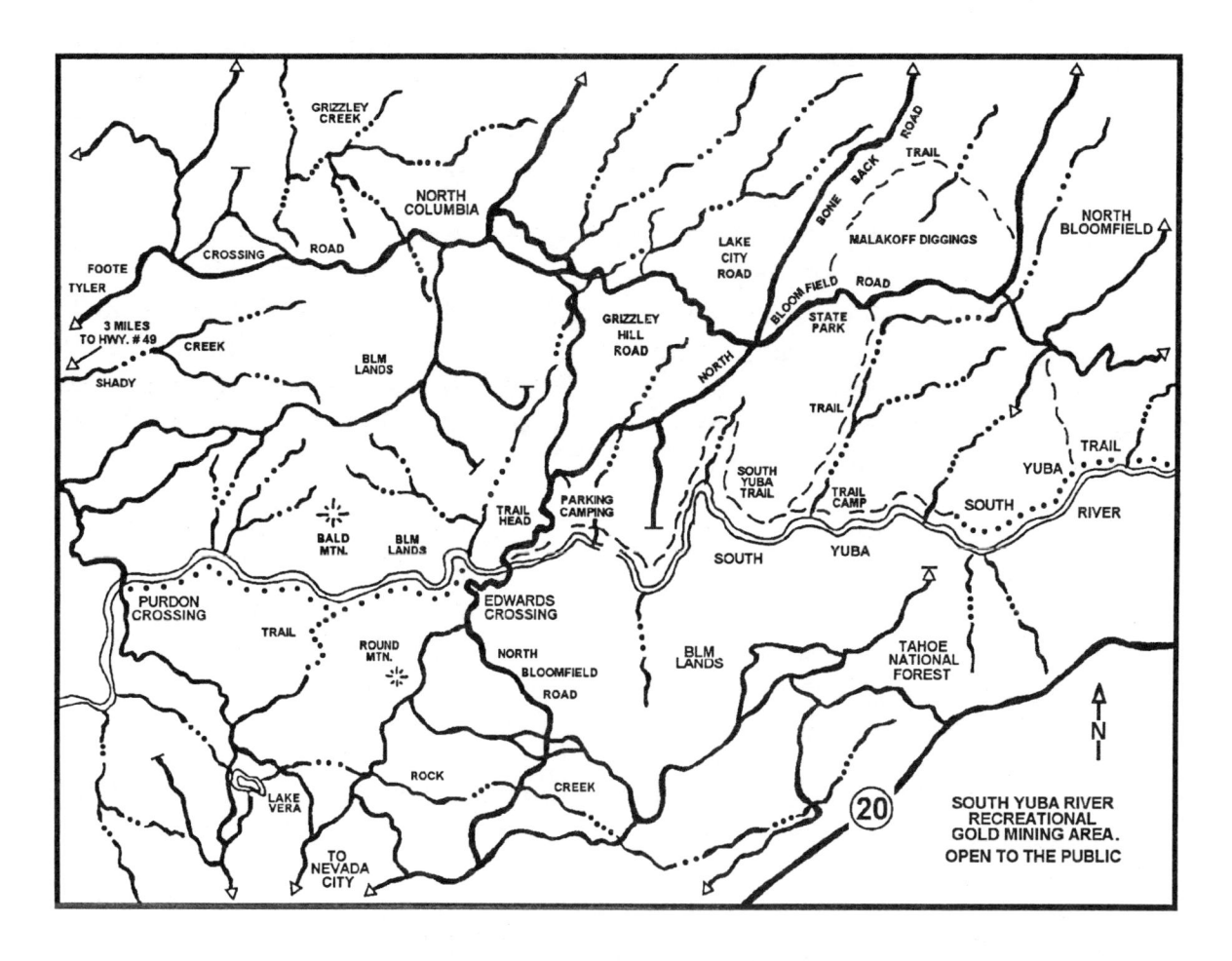

DEL NORTE COUNTY. CLASS E. Gold can be found along almost any water-run and their feeder creeks in DEL NORTE COUNTY. Open to dredging from July 1 through September 30, except where noted. The majority of the placer gold occurrences are located in the central portion of the County. Several areas are withdrawn from mineral entry and should be researched with the Six Rivers National Forest Office in Eureka, California.

The most favorable areas are at Gasquet Mountain, Low Divide, French Hill, High Plateau, Rattlesnake and Coon Mountain where gold placer, rare earth metals, platinum group metals and some *diamonds* have been located. Be alert to existing mining claims and private property. Check with County recorders office for legal access. Upper-levels of the Pliocene Age could be the extension of the placer gold deposits discovered in Josephine County, Oregon, where stringers play across the California land for hundreds of miles in the upper-elevation of graveled channels.

Craigs Creek T16N R1E Sec. 1 & 2, has expelled gold, platinum and chrome from its gravel benches, ancient terraces and basins. Placer gold mining was experienced on Hurdygurdy and Jones Creeks where adjacent benches lay from pre-Cretaceous Age. Big Flat (see page 61) is a vivid example of a large ancient graveled terrace laced with placer gold that lays between two creeks. Shelly, Patrick, Monkey Creeks and tributaries with their feeder creeks have seen placer gold mining that had produced considerable amounts of gold from its gravel. The many water-runs of DEL NORTE COUNTY continues to expel placer gold from its gravels. Check the tailing piles for overlooked gold 'n platinum nuggets.

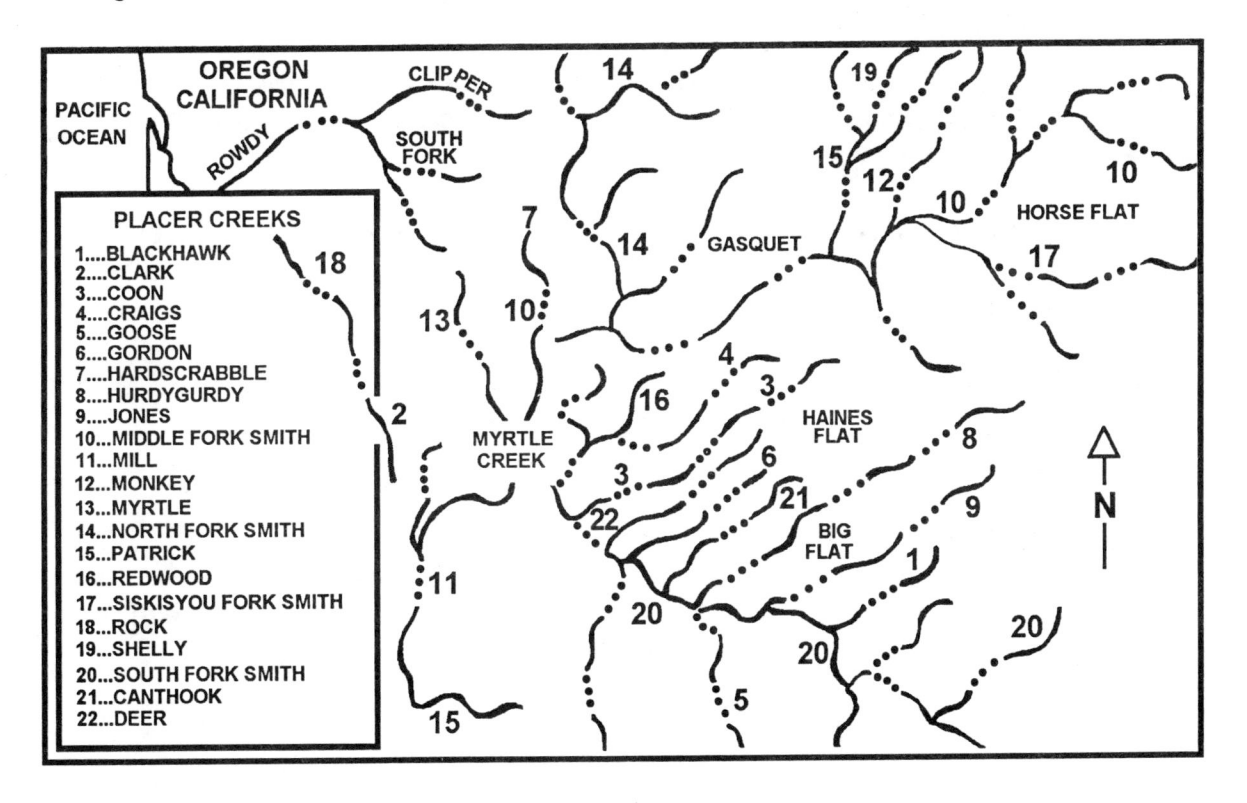

PLACER CREEKS
1....BLACKHAWK
2....CLARK
3....COON
4....CRAIGS
5....GOOSE
6....GORDON
7....HARDSCRABBLE
8....HURDYGURDY
9....JONES
10...MIDDLE FORK SMITH
11...MILL
12...MONKEY
13...MYRTLE
14...NORTH FORK SMITH
15...PATRICK
16...REDWOOD
17...SISKISYOU FORK SMITH
18...ROCK
19...SHELLY
20...SOUTH FORK SMITH
21...CANTHOOK
22...DEER

GOLD OCCURRENCES IN DEL NORTE COUNTY, CALIFORNIA.
SIX RIVERS NATIONAL FOREST. *Class - E.*

Many of the water-runs in Del Norte County give up a variety of grades in gold placer. A good gold panner can locate gold particles in the form of nuggets, flake, fines, flour and micron.

Much of the north-central area of the County is claimed staked at T17N R3E south of the Smith River, and T15N - T14N R2E south of Hardin Mountain. The remainder of the County is *open* to exploration, except for the areas set aside as State Parks, National Parks and the Siskiyou Wilderness Area.

Gold-bearing streams containing gravel, aggregate, sandbars, benches and high-banked terraces occur along many of the drainage's of Myrtle Creek, Jones Creek, Craigs Creek, Hurdygurdy, Patrick Creek and their many tributaries.

In the past, at higher elevations, water was a problem for the hydraulic miners to reach the deep gold placer gravel. *Much of this placer remains untouched and continues to re-supply the adjacent water-runs with placer gold.*

Geologic events laid down favorable gold occurrences in the many high-level gravels found to exist on capped Mountain tops, heavily forested plateaus and gravel terraces that were formed into mesa-type tablelands. Gold-bearing gravels were scattered across much of the Pliocene Aged formations that fingered down into Del Norte County from the rich gold-bearing deposits in Josephine County, Oregon.

To reach the back-country where the many gold-bearing water-runs and their tributaries are located, the following numbered Forestry Camp Sites can be used as base camps.

CEDAR RUSTIC (1), is located 31 miles NE of Crescent City on US Highway 199. With 12 gold sites accommodating 22 foot long trailers. Forestry camp sites are located at T14N R3E Sec. 16-NW1/4.

PATRICK CREEK (2), sets at 27 miles NE of Crescent City on US #199. With 17 camp sites accommodating 22 foot trailers. Forestry camp sites are located at T17N R3E Sec. 16-NW1/4.

SHELLY CREEK (3). Primitive camp site. Travel about two miles north from Patrick Creek campground onto Forestry Road # 316. Forestry camp sites are located at T17N R3E Sec. 5-NE1/2. Back-country water-runs are influenced by the extension of the rich gold deposits of Josephine County, Oregon.

GRASSY FLAT (4), is located 25 miles NE of Crescent City on US Highway 199. With 19 camp sites accommodating 22 foot trailers. Forestry camp sites are located at T17N R2E Sec. 24-NE1/4.

PANTHER FLAT (5), is located 22 miles NE of Crescent City on US Highway 199. With 42 camp sites accommodating 22 foot trailers and accommodations for the handicapped. Forestry camp sites are located at T17N R2E Sec. 27-NE1/4.

BIG FLAT (6), is located 11 miles NE of Crescent City on US Highway # 199 to County Road # 427, turn south, travel 13 miles to the BIG FLAT (6) campground. With 30 camp sites accommodating 22 foot trailers, located on the South Fork of the Smith River. Forestry camp sites are located at T15N R2E Sec. 23-NE1/4.

EARLY HOLE (7), is located 7 miles NW of Hiochi on the Smith River within the Jedediah Smith Redwood State Park. Forestry camp sites are located at T15N R1E Sec. 31-NW1/4. Use as a base camp for searching the back-country water-runs, creeks and their tributaries.

MILL CREEK (8), is located 7 miles south of Crescent City on the Redwood Highway US # 101 within the Del Norte Coast Redwoods State Park. Use these camp sites as a base camp for entering the back-country to explore and put down test holes for gold occurrence deposits. Forestry camp sites are located at T15N R1E Sec. 7-SE1/4.

TWO NO-NAME PRIMITIVE CAMPGROUNDS (9). Use the Six Rivers National Forestry Map for locations at T12N R4E Sec. 28. Located on Notice Creek at the junction of Bluff Creek. Second location one-half mile south of junction just below the ox-bow on Bluff Creek. *These are mineral withdrawal sites. Use as a base camp for serious gold searching the adjacent creeks and their tributaries.*

Leave Bluff Creek Ranger Station located on Highway # 96 and travel north on Forestry Road # 13N01 for about 13-plus miles to the junction of Forestry Road 12N010. Travel about four twisting miles to the confluence of Notice Creek and Bluff Creek. *Coarse gold and platinum* have been notably found to exist in the area bench gravels. Look for mini-gravel terraces that are located between creeks and streams. Be alert to existing mining claims and private property.

DELOS TOOLE ©

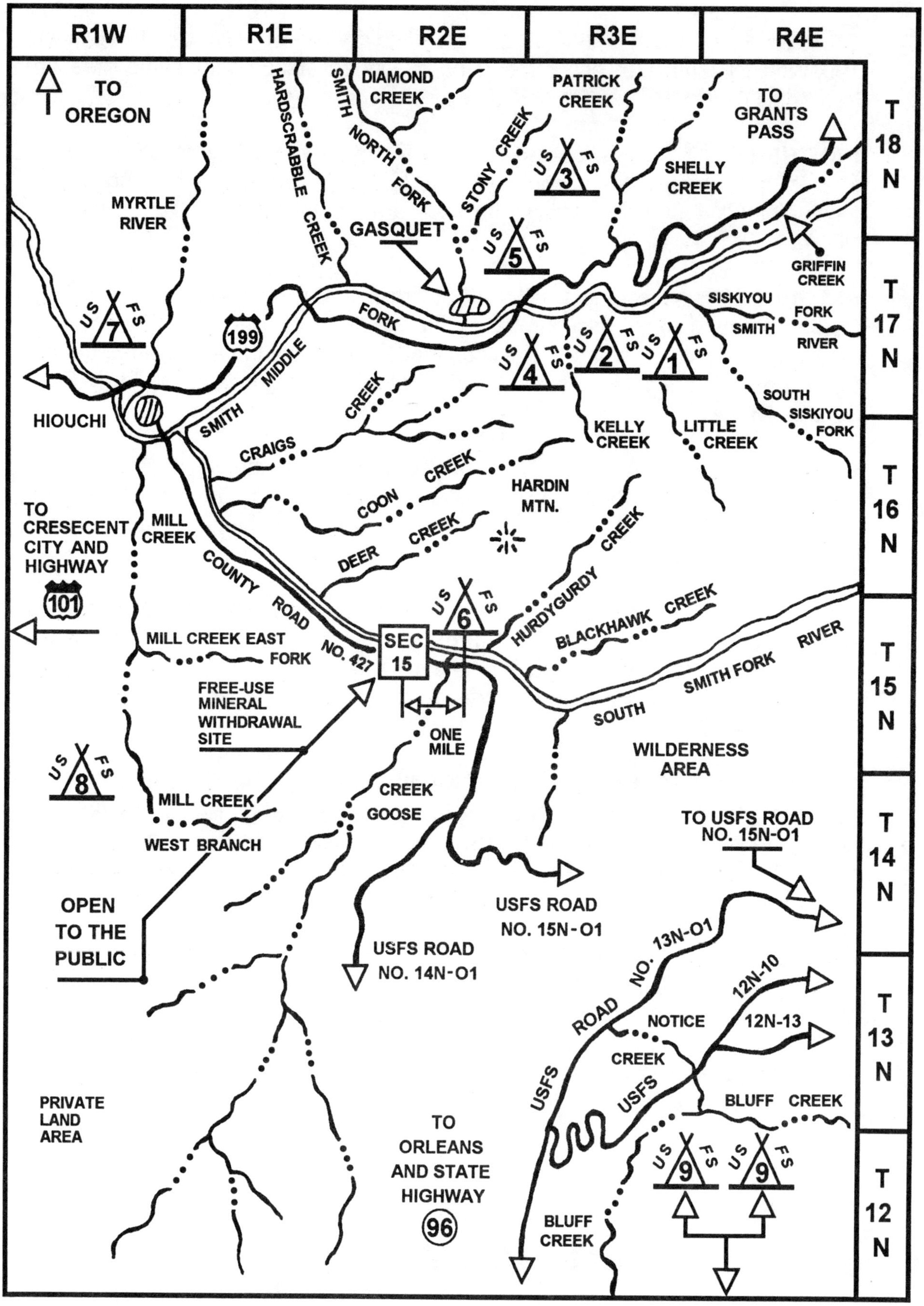
R1W
R1E
R2E
R3E
R4E
T 18 N
T 17 N
T 16 N
T 15 N
T 14 N
T 13 N
T 12 N
TO OREGON
MYRTLE RIVER
HARDSCRABBLE CREEK
SMITH NORTH FORK
DIAMOND CREEK
STONY CREEK
PATRICK CREEK
TO GRANTS PASS
SHELLY CREEK
GASQUET
GRIFFIN CREEK
SISKIYOU FORK
SMITH RIVER
SMITH MIDDLE FORK
199
HIOUCHI
CRAIGS CREEK
KELLY CREEK
LITTLE CREEK
SOUTH SISKIYOU FORK
COON CREEK
HARDIN MTN.
TO CRESECENT CITY AND HIGHWAY
101
MILL CREEK
DEER CREEK
HURDYGURDY CREEK
COUNTY ROAD NO. 427
MILL CREEK EAST FORK
SEC 15
BLACKHAWK CREEK
SOUTH SMITH FORK RIVER
FREE-USE MINERAL WITHDRAWAL SITE
ONE MILE
WILDERNESS AREA
MILL CREEK
WEST BRANCH
CREEK GOOSE
TO USFS ROAD NO. 15N-O1
OPEN TO THE PUBLIC
USFS ROAD NO. 15N-O1
USFS ROAD NO. 14N-O1
USFS ROAD NO. 13N-O1
12N-10
12N-13
NOTICE CREEK
BLUFF CREEK
PRIVATE LAND AREA
TO ORLEANS AND STATE HIGHWAY
96
BLUFF CREEK
US FS

HUMBOLDT COUNTY - SIX RIVERS NATIONAL FOREST WITH ACCESS TO THE BACK-COUNTRY.

AIKENS CREEK (10), the campground is situated at the confluence of Aikens Creek and Bluff Creek. With easy access from State Highway # 96 along the Klamath River. Smaller creeks in the area go dry during long hot spells, making gold sniping with a metal detector and dry washing activity easier to cover larger areas. Check those old tailing heaps and aging ore debris dumps left behind by the old miners. **Platinum** values were discarded in lieu of sought-after rich gold placers. Forestry camp sites are located at T10N R5E Sec. 30.

BLUFF CREEK (11), this campground is located just about one-mile east of the Aikens Creek campground and is situated along State Highway # 96, with location on the Bluff Creek proper ten-miles SW from Orleans. Use the area for a base camp for jumping off into the back-country in search of the placer gold graveled creeks. Forestry camp sites are located at T10N R5E Sec. 19.

PEARCH CREEK (12), is located one-mile NE of Orleans on State Highway 96 at the junction of Pearch Creek and the Klamath River. Use the camp sites for jumping off into the back-country when in search of gold-bearing creeks. Forestry camp sites are located at T11N R6E Sec. 32.

JUNCTION OF ADAMS CREEK WITH RED CAP CREEK (13), its location is four-miles SW of Pearch Creek by traveling Rattlesnake Ridge Road. Make a right SW turn at the junction of Forestry Road 8Q100 and travel about one-mile to the junction of Adams Creek with Red Cap Creek. This is a **Mineral Withdrawal Site** with near-by gold-bearing deposits in older bench gravels found at higher levels from present running creek channels. Be alert to existing mining claims and private property. Forestry camps are located at T10N R5E Sec. 14.

WHITEMORE CREEK (14), this is a primitive **Mineral Withdrawal Site** located one and one-half miles NE of Pearch Creek campground on State Highway # 96 at where Whitmore Creek makes its junction with the Klamath River. Be alert to existing mining claims and private property in the area as much hydraulic mining of the ancient gravels has occurred in the years past. Check those unclaimed tailing piles and disturbed gravel terraces for over-looked gold nuggets, gold particles and platinum values. Forestry camps are located at T11N R6E Sec. 20.

ISHI PISHI ROAD (15), is a primitive **Mineral Withdrawal Site** located on the west opposite side of the Klamath River and State Highway # 96. One-mile NE of Pearch Creek camp site on the Ishi Pishi Road at location T11N R6E Sec. 29.

KLAMATH RIVER hosts a number of important tributaries and their feeder streams. The water-runs that junction with the Klamath River, just to name a few are, Scott, Salmon, Shasta River, Cottonwood, Seiad, Horse, Thompson, Clear, Indian, Dillon and Camp Creeks.

The Trinity River joins the Klamath River as a confluence at Weitchpec on the Hoopa Indian Reservation thirteen miles north of **Tish Tang (16) Campground.** The main tributaries of the Trinity River are, Stewart's Fork, East Fork, Coffee Creek, New River, Indian Creek and Hayfork Creek. Feeder streams carry the seasonal run-off of gold particles into the main channels. Some of these are the Sharber, China, Quinby, Coon, Kirkham, Cedar, Kahala, Ammon, Madden, Boise, Gray, Hennessy, North Fork Mingo & Hawkins Creeks.

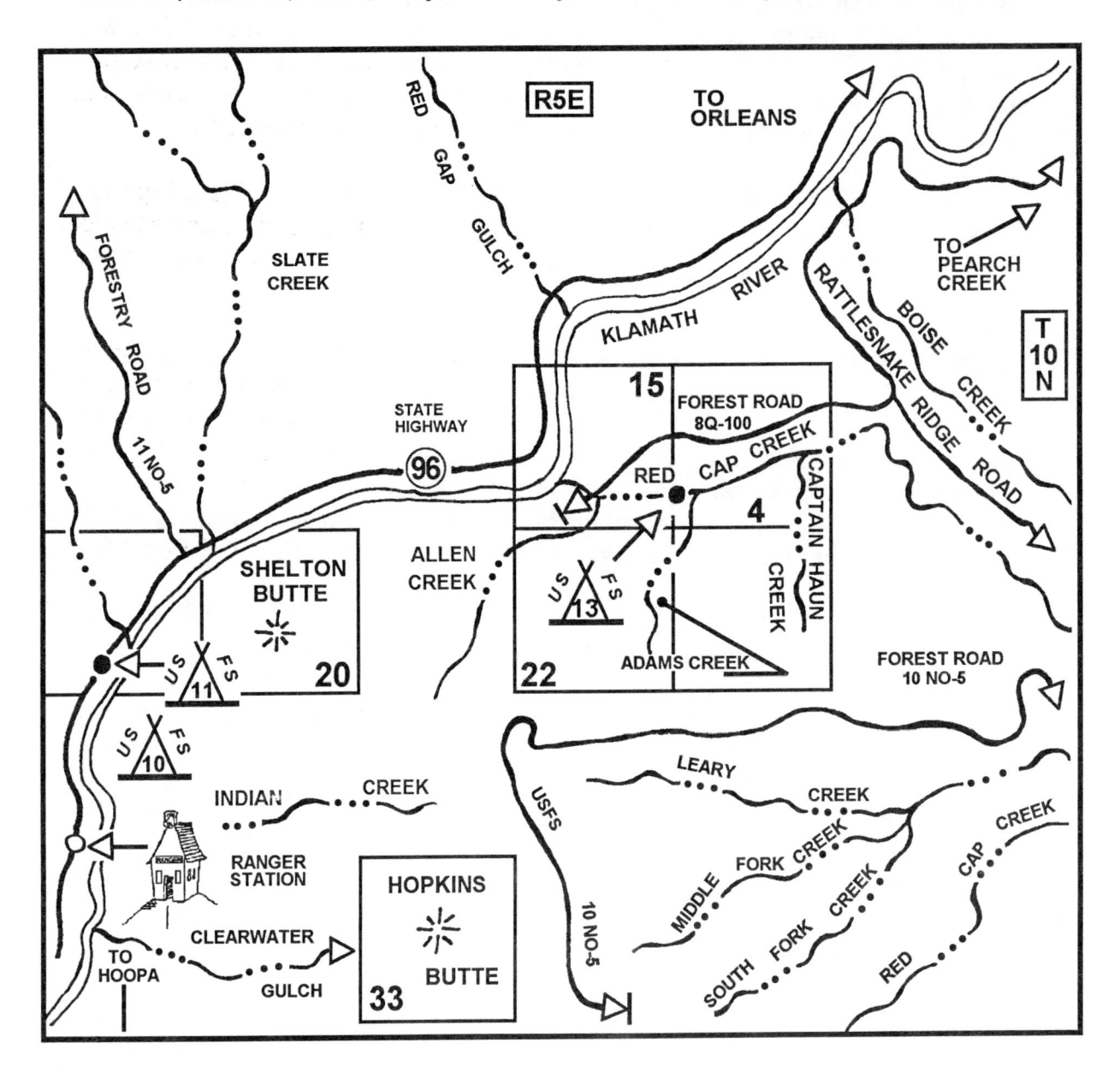

ORLEANS has had a history in placer operations in and around its perimeter with hydraulic mining being the main activity a few years ago. The Pearch hydraulic mine was one of Orleans principle mining operations.

The Klamath River and its many tributaries host stream gravels with extensive older bench aggregate laying at a higher level than the present flow of the water-runs. The gold is flour, fine to medium with evidence of some platinum being dredged. Hand wash the micron gold several times over to bring it into view. Check those tailing piles for over-looked *gold nuggets, platinum, rare earth minerals* and semi-precious gem stone material. Much of the *gold occurrence* was lost to the ore heaps due to greed that hastened the operations and let the *gold particles* to slip through the sluice.

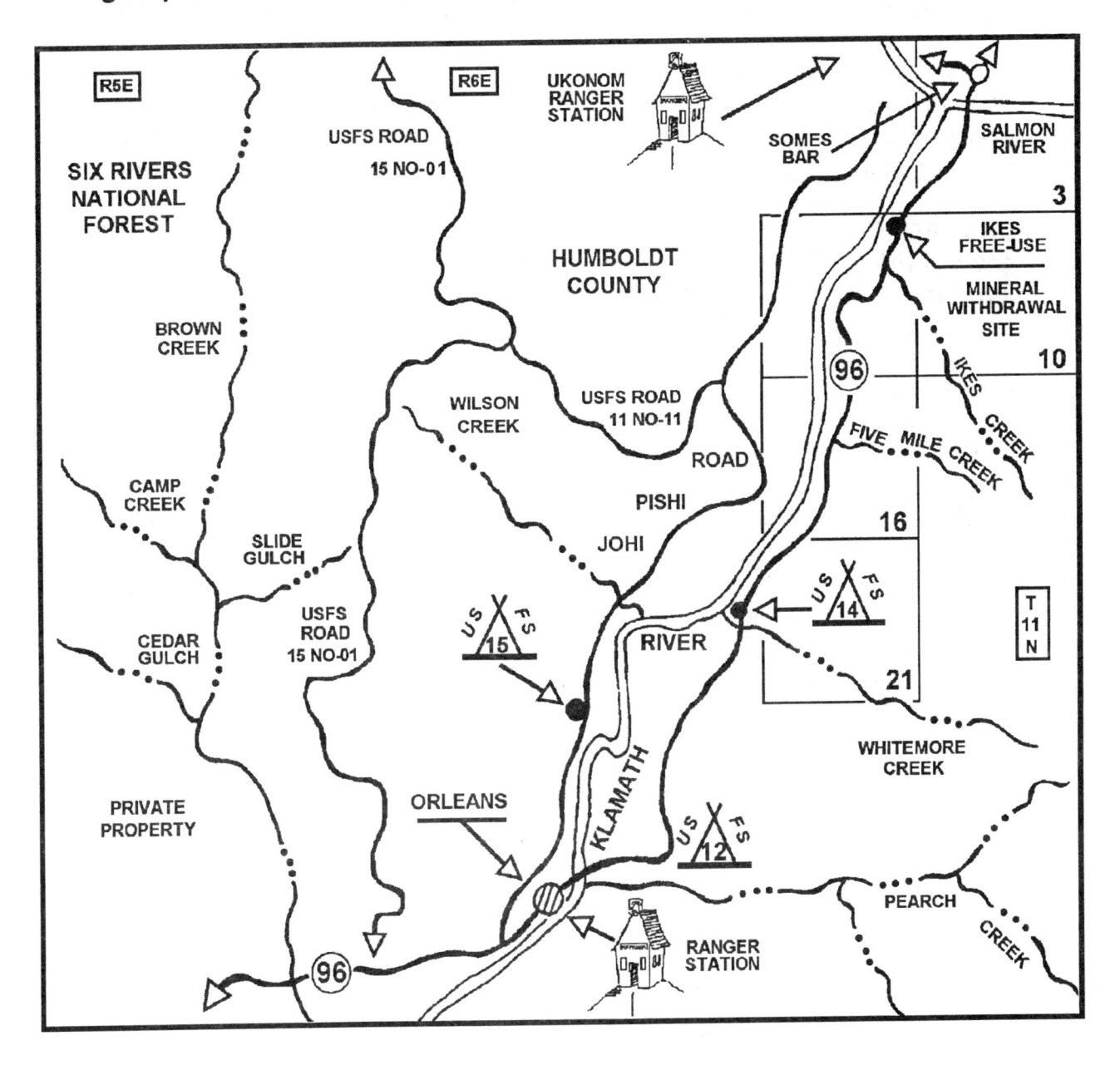

TISH TANG (16), campground is located on the bend of the Trinity River about two and one-half miles south from Hoopa, on State Highway # 96. All of the Forestry campgrounds, (except where noted), are Mineral Withdrawal Sites where *recreational gold panning is open as a Free-Use activity* but is closed to mineral claim staking. Equipment is limited to gold panning and with small hand held tools. Contact the Forest Ranger for additional information before beginning recreational mining in the area of interest.

Use Tish Tang site for base camp site while entering the back-country to explore for the *gold-bearing water-runs* and older terrace deposits along the banks of those feeder creeks. Forestry camp site is located at T7N R5E Sec. 5. Hoopa Indian Reservation is closed to mineral exploration and recreational gold mining.

BOISE CREEK (17) campground is located two miles SW of Willow Creek on State Road # 299 at the confluence's of Boise Creek, Brannon Creek and Willow Creek. *Use this camp site as a base camp for serious gold prospecting* in the back-country water-runs. Forestry camp site is located at T7N R5E Sec. 30.

EAST FORK (18) campground is situated on State Road # 299 about five and one-half miles SW of Boise Creek campground and located on Willow Creek at the junction of the East Fork Creek. Just off of the accompanied map the Forestry camp site is located at T6N R4E Sec. 15.

GRAY FALLS (19) campground is located on the Trinity River in Trinity County, and situated on State Highway # 299 about six miles from Salyer. The area has a rich history of gold placer mining, especially from the Trinity River and its tributaries. *The Trinity River is flanked by accessible back-country hosting gold bearing bench gravels.* Forestry camp site is located at T6N R6E Sec. 34.

MAD RIVER campground is located on County Road # 502 about five miles SE of Mad River (hamlet) and three miles north of the dam. This camp site exposes the gold searcher to the back-country for serious exploration. Be alert to mining claims and private property. The Forestry camp site is located at T1S R6E Sec. 2.

Use the Six-Rivers National Forestry Map for locating the listed camping sites. For further information contact Six-Rivers National Forest District Office at; 1330 Bayshore Way, Eureka, CA. 95501. 707-442-1721. Gasquet Ranger District, P.O. Box 228, Gasquet, CA., 95543. 707-457-3131. Orleans Ranger District, Drawer B, Orleans, CA., 95556. 916-627-3291. Lower Trinity Ranger District, P.O. Box 68, Willow Creek, CA., 95573. 916-622-2118.

NEW RIVER has been an important tributary of the Trinity River and is still regarded as a gold placer producer. New River flows through the Denny area and eventually joins up with the Trinity River at location where the Burnt Ranch camp site sets. *(Suggestion; go all the way up New River to the top of Battle Creek.)*

The area has witnessed hydraulic mining of the bench gravels with evidence of where undercutting of the (high-banking) auriferous gravel banks had taken place. Beneath the overgrowth lay placer deposits that have not been discovered because of the moratorium placed on hydraulicing practices a few decades ago. Be alert to existing mining claims and private property of the area when sniping with a metal detector for the overlooked gold nuggets that lay on the tailing piles. Good area to explore for dredging action.

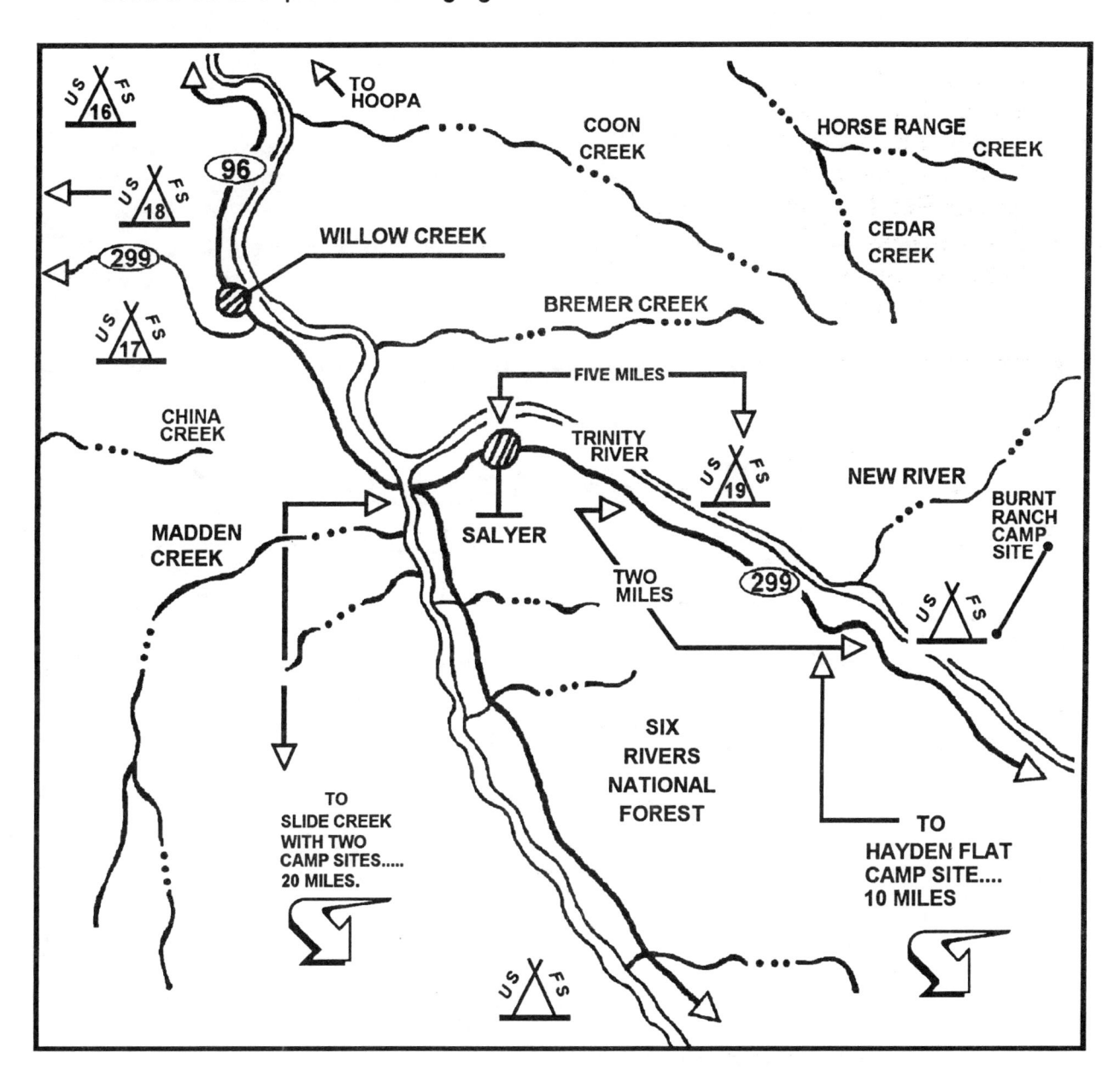

BOULDER CREEK and UPPER SACRAMENTO RIVER AT GIBSON ROAD.
Use maps for USGS Topographic *Chicken Hawk* at map location T36N R5W Sec. 35 SE1/4. Use Shasta National Forest Map, California Highway, Shasta County Map and the Automobile Club of Southern of California (AAA) Map, Located just off of I-5 two miles north of the Forestry Pollard Flat campground, sixteen miles south of Dunsmuir and thirty-five miles north of Redding, California.

After every rain and spring run-off, the area is replenished from the scattered gold deposits of the back-country, and from those ledges with narrow quartz veins in greenstone hosting free gold and minute amounts of sulfides. Placer in the area comes from older bench gravels and aggregate of the Cretaceous period. Look the area over carefully because the Sacramento River has changed its course several times in this particular area. The 1880 RR construction crews diverted the Sacramento River several times for the sake of laying RR track rails on more stable ground. This created stretches where the old river channel existed that became exposed and now can be uncovered and worked when located. Be alert to old placer deposits mixed in hard pack clay benches; remove flooded debris and check the layered sand; work the bridges where the I-5, Gibson Road and the RR tracks cross Boulder Creek with their gold catchers. Some places are better than others as the spring run-off moves area spots. Seasons change drop points of the gold and cover collecting dips with compacted sand.

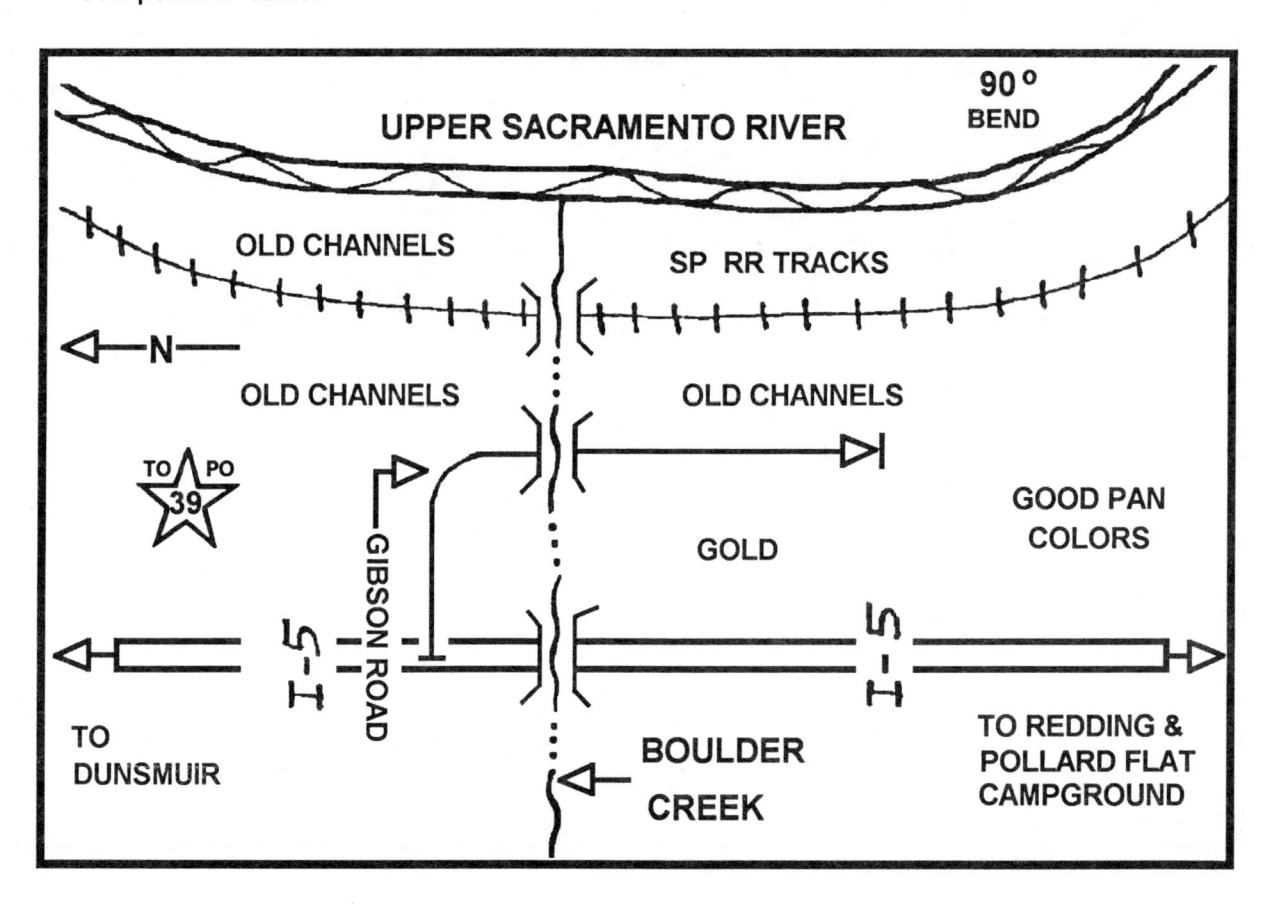

EL DORADO COUNTY. *East of Highway # 49 is CLASS C,* open to dredging from the fourth Saturday in May thru October 15, the remainder of the county is *CLASS H.* Open all year around. Use El Dorado National Forest, County Maps for location.

The ancestral Consumnes River crossed into the central portion of El Dorado County where crystalline limestone lay exposed to weathering, erosion, creating potholes, crevices and basins that attracted much placer **gold-bearing** gravel to settle in. Later, the modern day rivers cut through the holding depressions, scattering placer gold into adjacent water-runs. Those holding depressions continue to exist in overlooked and undiscovered locations in the back-country. Many USFS camp sites can be used as a base camp while searching the back-country for the gold occurrences. Adjacent to the Volcanoville east area, Tertiary quartz-rich **gold-bearing** gravels had been placer mined in the past. Present day eroding attacks the banks, terraces and gravel ledges, carrying the **placer gold** into near-by riffles of the present day American River and its tributaries.

The Tertiary gravel patches of the Georgetown area contain many exposed seam veins and deposits that feed the adjacent creeks with **placer gold** originating from the Sierra Nevada East Gold Belt and its foothills. Much of this modern **gold-bearing** influence moved into the rivers and creeks of Rubicon (T9N R12E0, Weber (T10N R10E), Rock (T11N R11E), Camp (T9N R12E) creeks, along with many other water-runs. The County had a very active past history in placer mining, lode and hydraulic mining. Check those tailing piles, old mine ore debris dumps with a metal detector for overlooked **gold nuggets**.

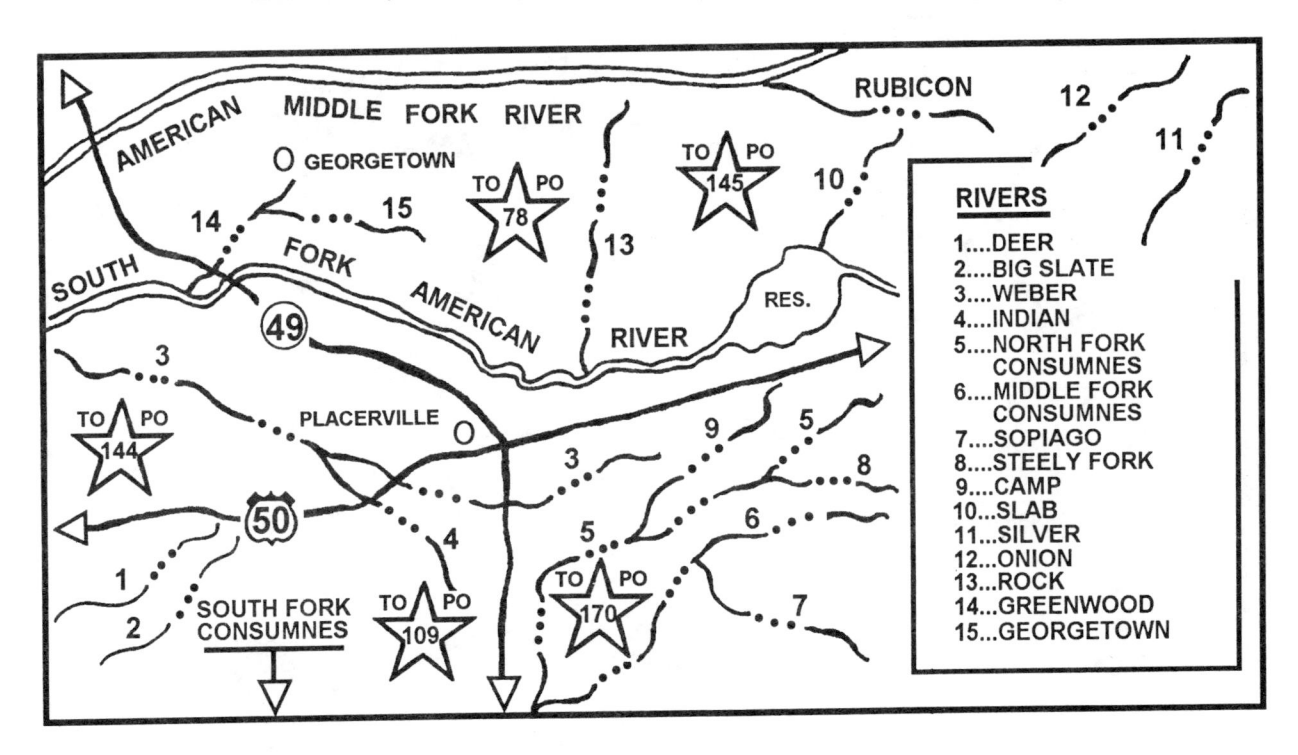

RIVERS

1....DEER
2....BIG SLATE
3....WEBER
4....INDIAN
5....NORTH FORK CONSUMNES
6....MIDDLE FORK CONSUMNES
7....SOPIAGO
8....STEELY FORK
9....CAMP
10...SLAB
11...SILVER
12...ONION
13...ROCK
14...GREENWOOD
15...GEORGETOWN

WEBER CREEK (EL DORADO COUNTY). From Highway # 50 crossing, east upstream, is *CLASS C.* Open to dredging from the fourth Saturday in May thru October 15.

EL DORADO COUNTY; east of Highway # 49 is *CLASS C.* Open to dredging from the fourth Saturday in May thru October 15. The remainder of El Dorado County is *CLASS H.* Open all year around except where noted.

Use El Dorado County Map, California Highway Map and the El Dorado National Forest Map for location. Be alert to existing mining claims and private property; check with the County Recorders Office for accessiblity into the area of interest.

WEBER CREEK bridge crossing at Highway # 50 is located two miles southwest of Placerville. The location can be found on the *PLACERVILLE TOPOGRAPHIC QUADRANGLE* at T10N R10E Sec. 14 and on the El Dorado National Forestry Map.

The area has had a history of rich placer mining activity that gave notoriety to the "Deep Blue Lead" channel that runs thru the west-southwest Placerville district. Gold is found in a wide area of the Quadrangle and comes from the seasonal run-off of the Mother Lode Belt gravel influences. The serpentine belt of the area extends itself into the south sector of the town of Placerville, where its exposed ledges of massive quartz veins enrich Weber Creek and its Tributaries with the **free-running gold particles.**

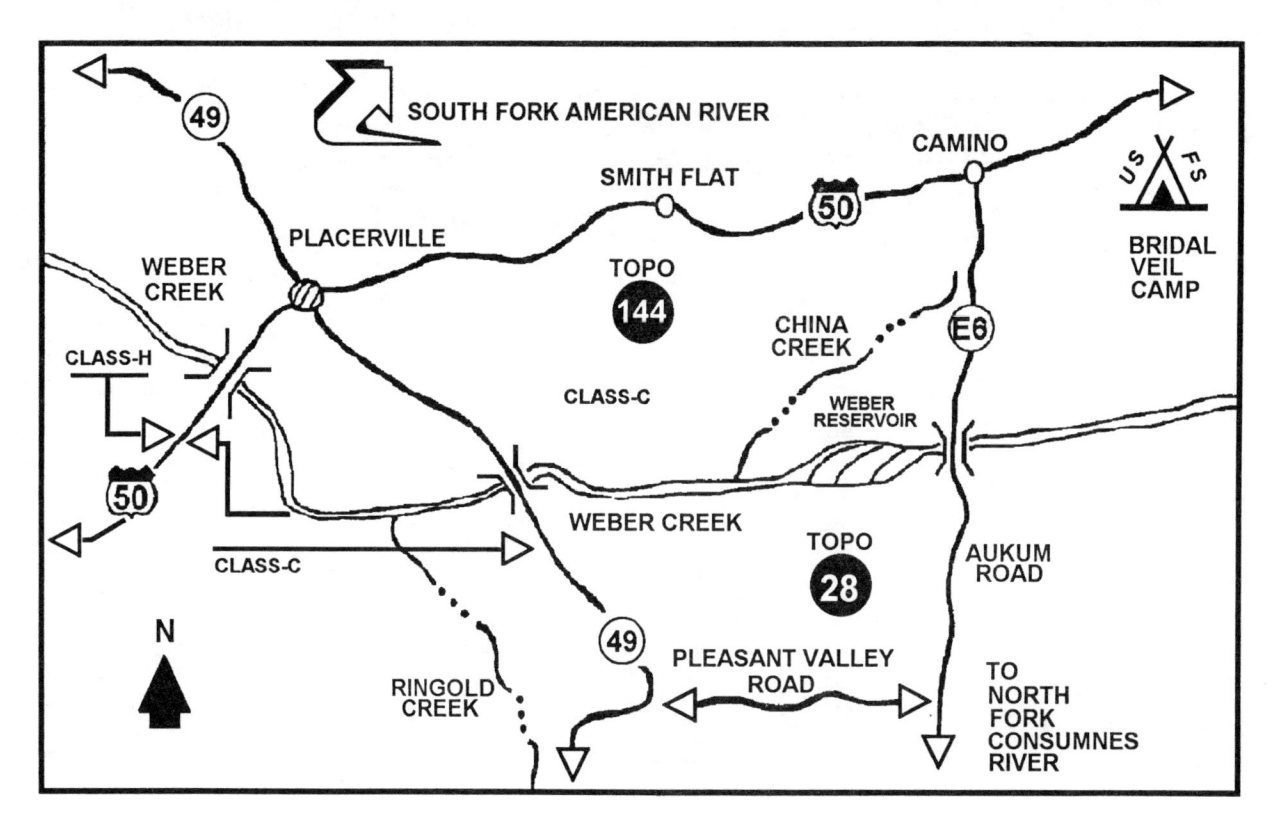

AMADOR COUNTY, east of Highway # 49 is **CLASS C.** Open to dredging from the fourth Saturday in May thru Oct. 15. Remainder of the County is **CLASS H.** Open all year. The northern border of the County is marked by the Consumnes River and its South Fork. The southern border is lined with the Mokelumne River with several reservoirs and the Mokelumne North Fork River. In between the north and south County line are hundreds of feeder creeks that have been influenced by the many intermittent basins of quartzite gravels, where *placer gold* was deposited in the channels by the ancient Tertiary Consumnes River.

There are narrow gold-quartz stringers spider-webbed thrue the section of Amador County, starting from the old mining town of Fiddletown (T8N R11E), to the south near Jackson (T6N R11E). Feeder creeks adjacent to the old mines are seasonally re-supplied with *gold particles* from the run-off of the old mine debris dumps and tailing piles.

Notable dredging occurred with rich results of the past in many of the water-runs north of Ione(T6N R9E), with Mule and Horse Creeks being worked several times over. Eons of activity from the forces of erosion concentrated *gold particles* in small, but rich pockets near Volcano (T6N R12E), that gave this area a historical background of having been the richest placer diggings in California. A sliver of the ancient channel appears near Pine Grove where the adjacent feeder creeks are influenced with *placer gold* from this underlying gravel mass during seasonal run-off, (T6N R12E). Be alert to mining claims and private property.

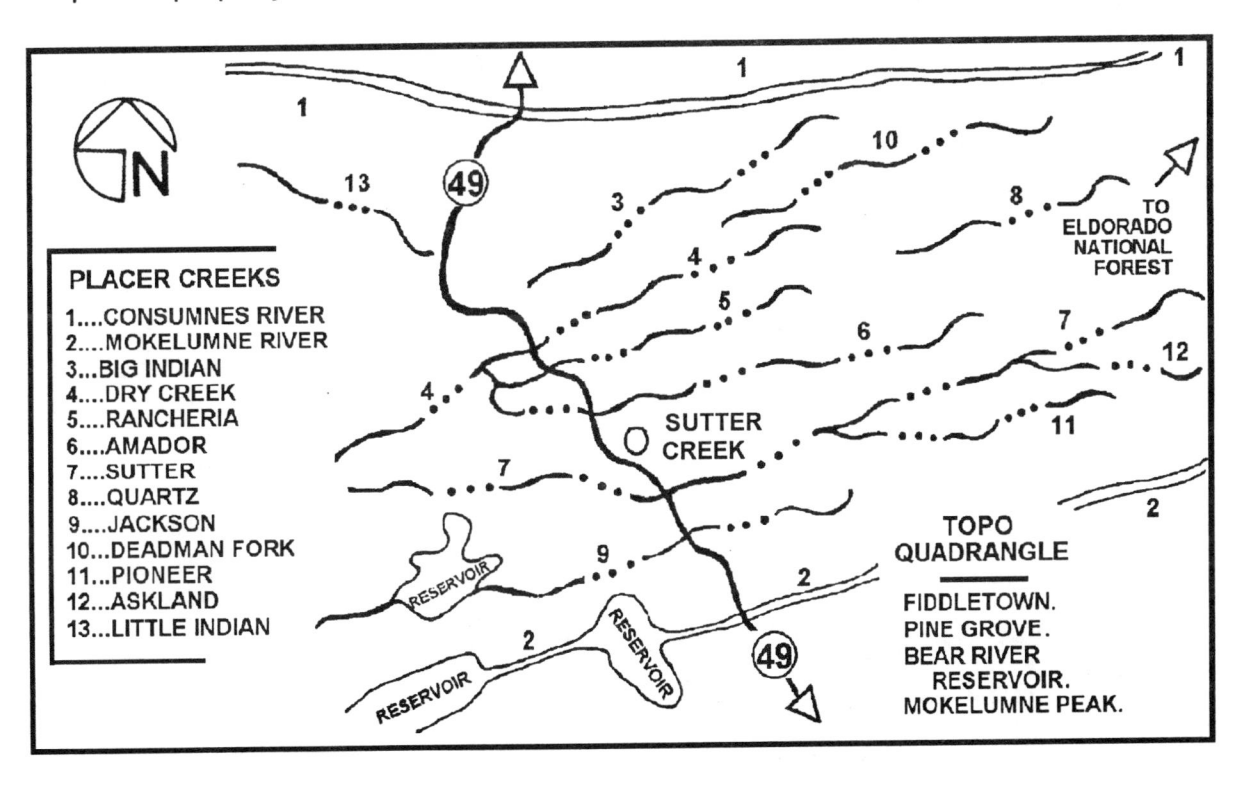

PLACER CREEKS
1....CONSUMNES RIVER
2....MOKELUMNE RIVER
3....BIG INDIAN
4....DRY CREEK
5....RANCHERIA
6....AMADOR
7....SUTTER
8....QUARTZ
9....JACKSON
10...DEADMAN FORK
11...PIONEER
12...ASKLAND
13...LITTLE INDIAN

TOPO QUADRANGLE

FIDDLETOWN.
PINE GROVE.
BEAR RIVER RESERVOIR.
MOKELUMNE PEAK.

CONSUMNES RIVER (SACRAMENTO, AMADOR and EL DORADO COUNTIES).

From the Western Pacific Railroad Bridge, (just about one-mile east of Highway I-5 and 1/4 mile east of the confluence with the Consumnes River and the Mukelumne River, in the Sacramento County), east, upstream to the Latrobe Highway Bridge in El Dorado County, is **CLASS D**. Open to dredging from July 1 thru September 15 and its tributaries inclusive.

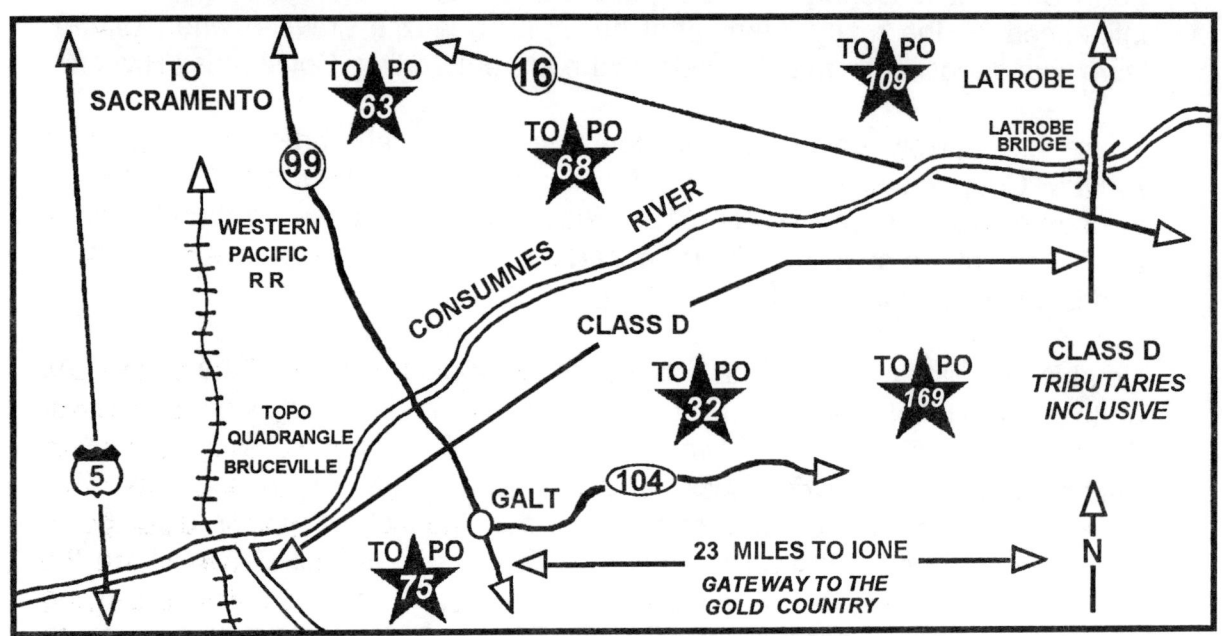

From the Latrobe Highway Bridge, east, upstream of the Consumnes River, to the State Highway # 49 is **CLASS H**. North Fork Consumnes River from its junction with the Consumnes South Fork River upstream to the bridge on County Road # E-16 is **CLASS H**. Middle Fork Consumnes River from its junction with the Consumnes South Fork River, upstream to the County Road Bridge # E-16 is **CLASS H**.

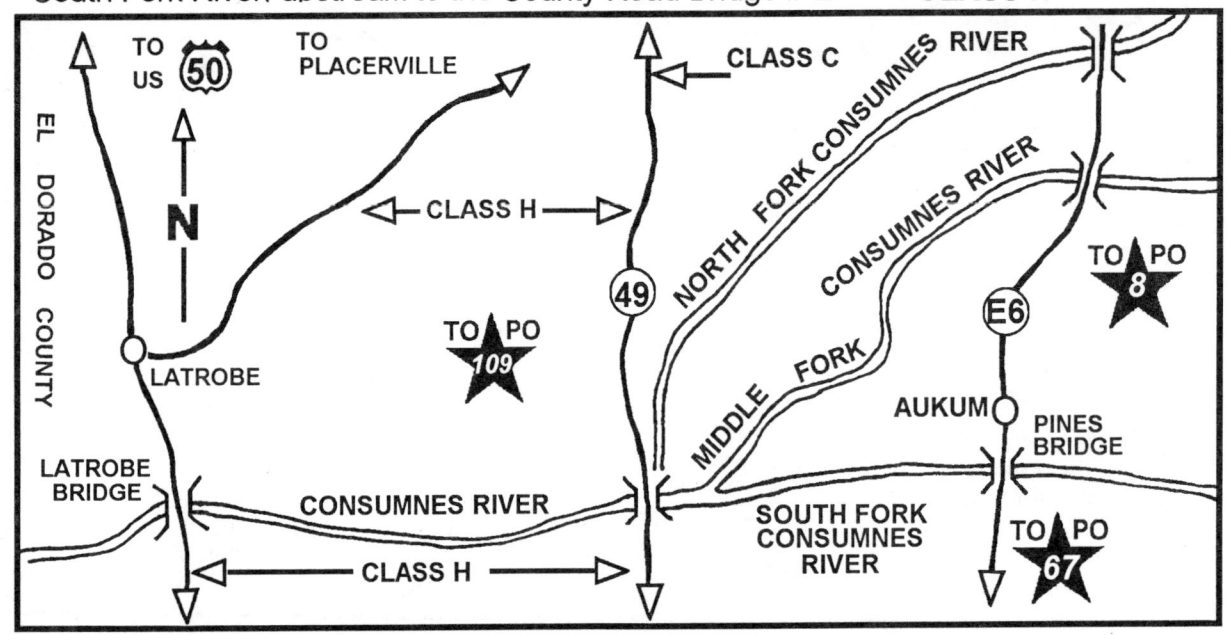

RUBICON RIVER (PLACER and EL DORADO COUNTIES). From its junction with the Middle Fork of the American River, east, upstream to the Georgetown Divide-Ralston Ridge Road Crossing at location, T13N R12E Sec. 7 is **CLASS C.** From its junction with the Middle Fork of the American River, east, upstream to Hell Hole Dam, including their tributaries, no dredge with an intake larger than **FOUR INCHES** may be used.

PLACER and EL DORADO COUNTIES, east of Highway # 49 is **CLASS C,** open to dredging from the fourth Saturday in May thru October 15. The remainder of the County is **CLASS H.** Open all year around except where noted.

Many patches of auriferous Tertiary gravels were worked by hydraulic methods in the past. Check those tailing piles 'n heaps. A favorite area with dredgers and gold snipers searching for **gold nuggets** with their metal detectors. Be alert to existing mining claims and private property. Check with the County Recorders Office for accessibility in the area of interest.

The back-country hosts many hills with exposed "seam" gravel basin deposits that is found to hold Tertiary **gold pockets** with scattered progenitors of quartz veins in finely disseminated particles of gold and pyrite. Heavy rains of the season rake 'n wear those **gold exposures** into lower water-run riffles and provide the dredgers and snipers with a ready made "horn of plenty". The north end of the northeast area of the Mother Lode Belt influence extends through this portion of the "Old Blue Lead", carrying its spider-webbed influence thru ancient channels of sixty-million year old aggregate gravels.

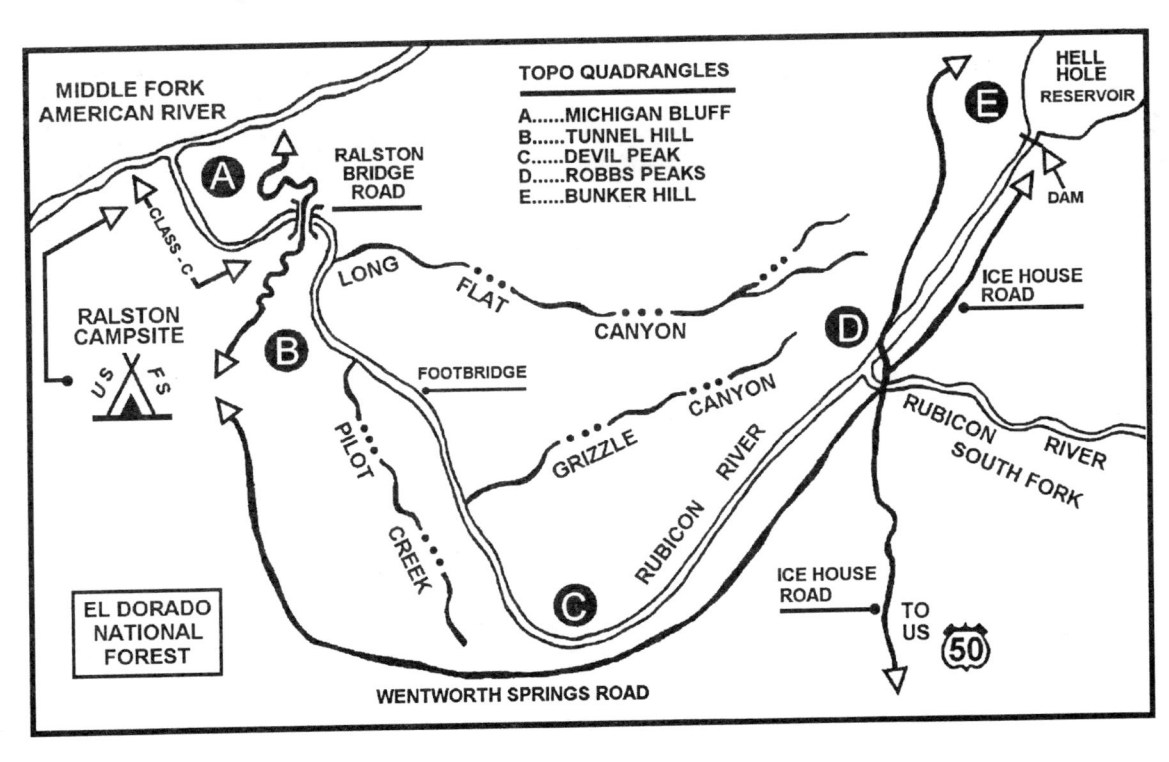

TOPO QUADRANGLES

A......MICHIGAN BLUFF
B......TUNNEL HILL
C......DEVIL PEAK
D......ROBBS PEAKS
E......BUNKER HILL

AMERICAN RIVER, SOUTH FORK (EL DORADO COUNTY). From Folsom Reservoir upstream to Highway # 49 bridge at Coloma is *CLASS - C.* Use the El Dorado National Forest Map, California Highway and the El Dorado County Maps for location.

The South Fork American River at this location is heavily influenced by the ancient tributary of the Tertiary Channel known to host massive quartz veins with hundreds of parallel stringers of disseminated free-gold that feed the many tributaries with its *flood gold particles and placer gravel gold.* The Tertiary gravels release its gold during the winter run-offs and spring time thunder storm flash flooding into the area feeder water-runs.

For a complete list of all of the State Park laws governing human behavior under Title 14, write to; Department of Parks and Recreation, 1416 9th. St., Box 42896, Sacramento, CA., 94296-0001. Five primitive campgrounds are available in the Recreation Area with a 30 day total camping limit in any calendar year in the park. For further information about campgrounds, write to; American River District, California Parks and Recreation, 7806 Folsom-Auburn Road, Folsom, CA., 95630. 916-988-0205.

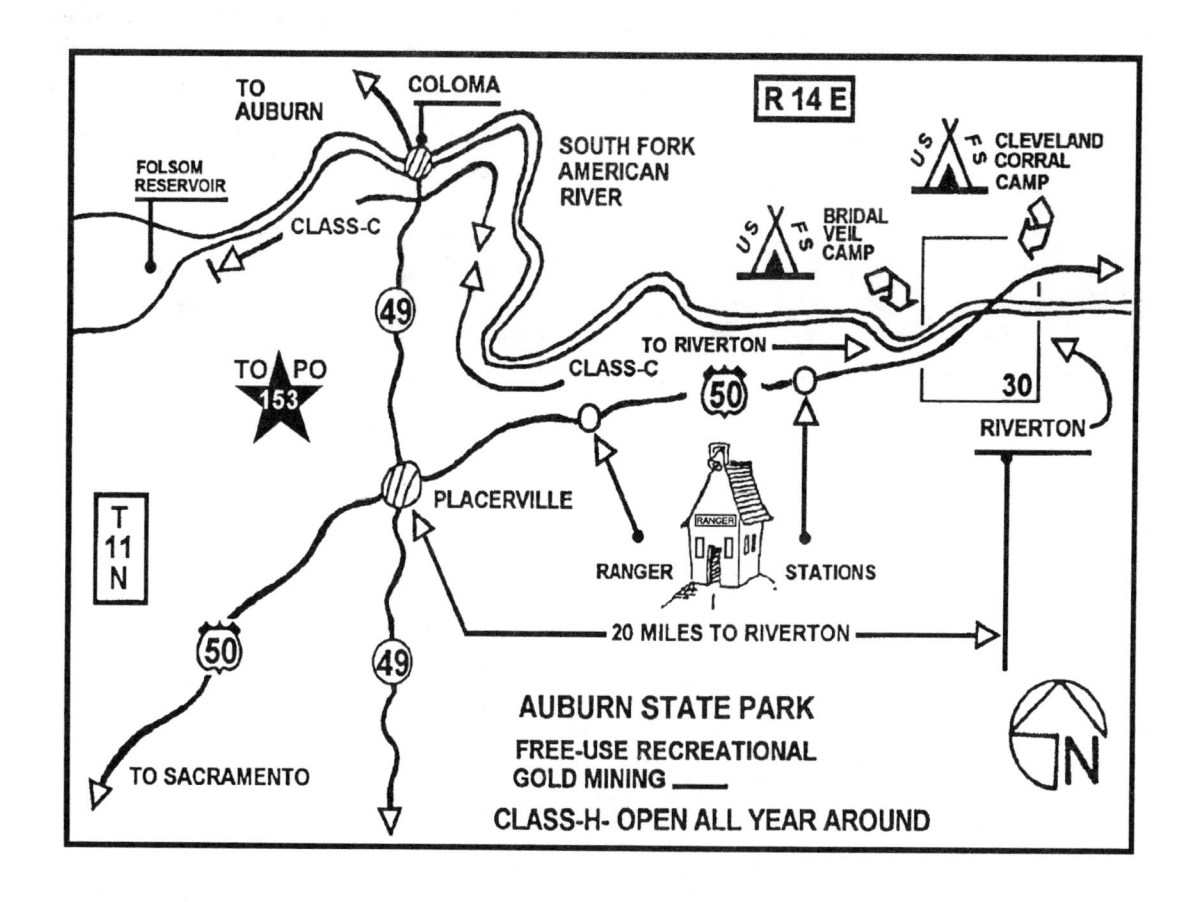

AMERICAN RIVER SOUTH FORK *TRIBUTARIES* (EL DORADO COUNTY).
All tributaries to the South Fork of the American River from the Folsom Reservoir to the Chili Bar Bridge, at location T11N R10E Sec. 35, is *CLASS - C,* open from the fourth Saturday in May thru October 15 for vacuum-suction dredge mining activity. (For dredging permit/license info. write or call; Calif. Fish 'n Game License 'n Revenue Branch 3211 "S" St., Sacramento CA 95816   916-327-0195.)

Tributaries within this area are; Indian Creek, Blue Tent Creek, Black Rock Creek, Greenwood Creek, North Ravin Creek, Burnt Shanty Creek, Granite Creek, Cold Creek, Kelsey Canyon, Dutch Creek, Chuck Ray Creek, Brush Creek, Big Canyon, Hangtown Creek, White Rock Canyon and Weber Creek (North and South Fork).

These tributaries all have feeder-creeks that refurbish their gold particle content from the adjacent ancient gravel terraces, benches and channels that contain Tertiary gold. Much of the countryside is made up of deep old river channels that host rich placer gold. Winter rains, winter run-off and summer run-off rake the exposed gravel banks sending the loosened gold into neighboring creeks to settle into the riffles best suited for catch all basins. **The area is host to many old hydraulic tailing piles 'n heaps.**

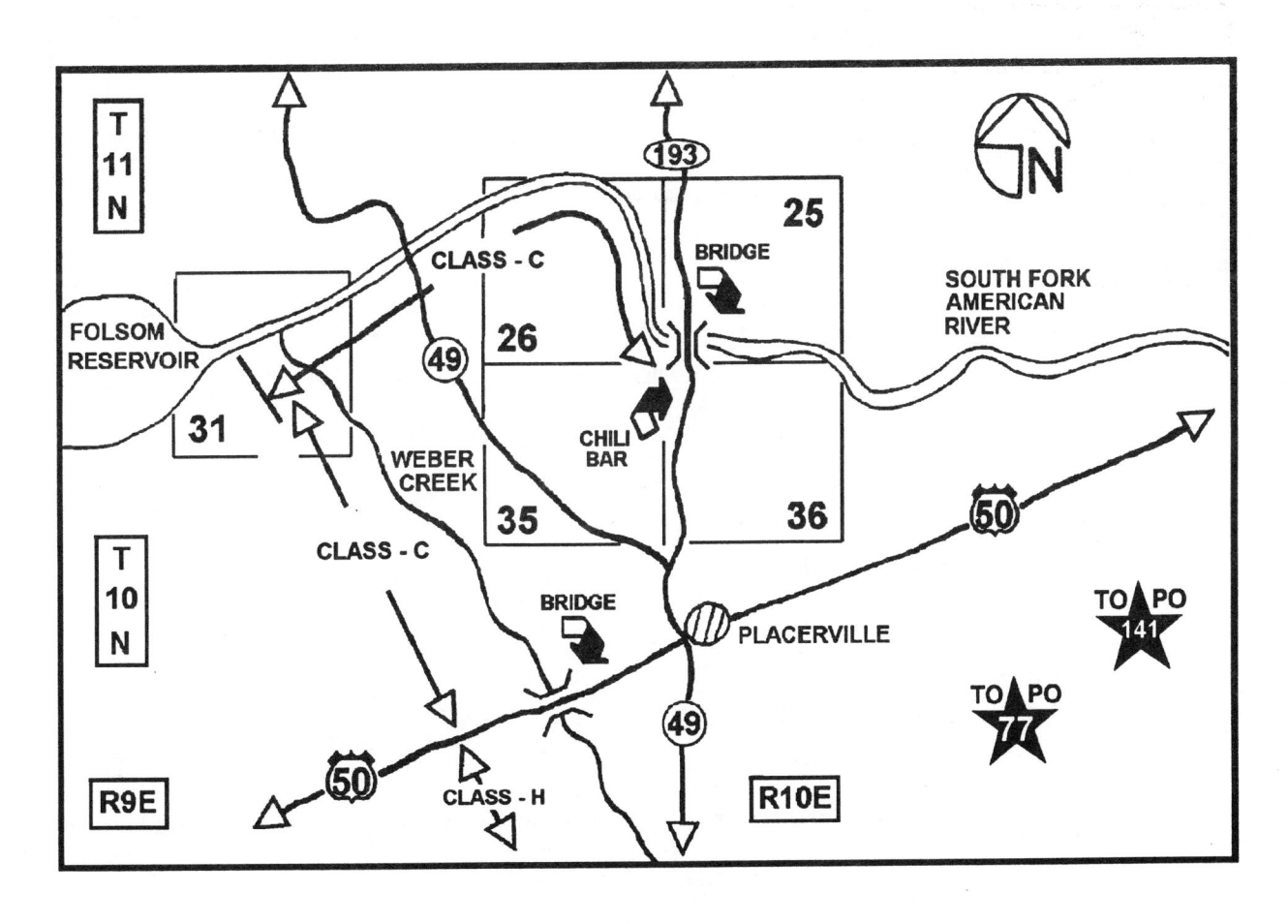

AMERICAN RIVER NORTH FORK (EL DORADO & PLACER COUNTIES).
From Folsom Reservoir to 1,000 feet upstream from Colfax Iowa Hill Bridge is *CLASS - C.* From 1,000 feet upstream from Colfax Iowa Hill Bridge to Heath Springs at location T16N R14E Sec. 26, is *CLASS - A.* **OPEN TO THE PUBLIC AS A FREE-USE RECREATIONAL GOLD MINING SITE AND THE RECREATION LAND PROVIDING A VARIETY OF USES. THE AUBURN STATE RECREATION AREA** is located along more than 30 miles and includes the North and Middle Forks of the American River with over 30,000 acres where the recreational gold miner can search for gold. A valid dredging permit from the California Department of Fish and Game is required for active dredgers and their helpers. Filing a mining claim within the Park is prohibited. Recreational gold panning, dredging and rockhounding is allowed in all of the permanent running stream-beds in the Auburn State Recreation Area of the Park. All regulations, rules of the land and restrictions must be observed when panning for gold, using a suction-vacuum dredge, using a metal detector for gold sniping and when in the back-country searching for the gold occurrences and gold particles. The area has had a rich history of gold production from its feeder streams, gravel benches and ancient fluvial plains. For further information about recreational gold mining, write; California Department of Parks and Recreation, Auburn State Recreation Area, 7806 Folsom Auburn Rd. Folsom, CA 95630   916-988-0205.

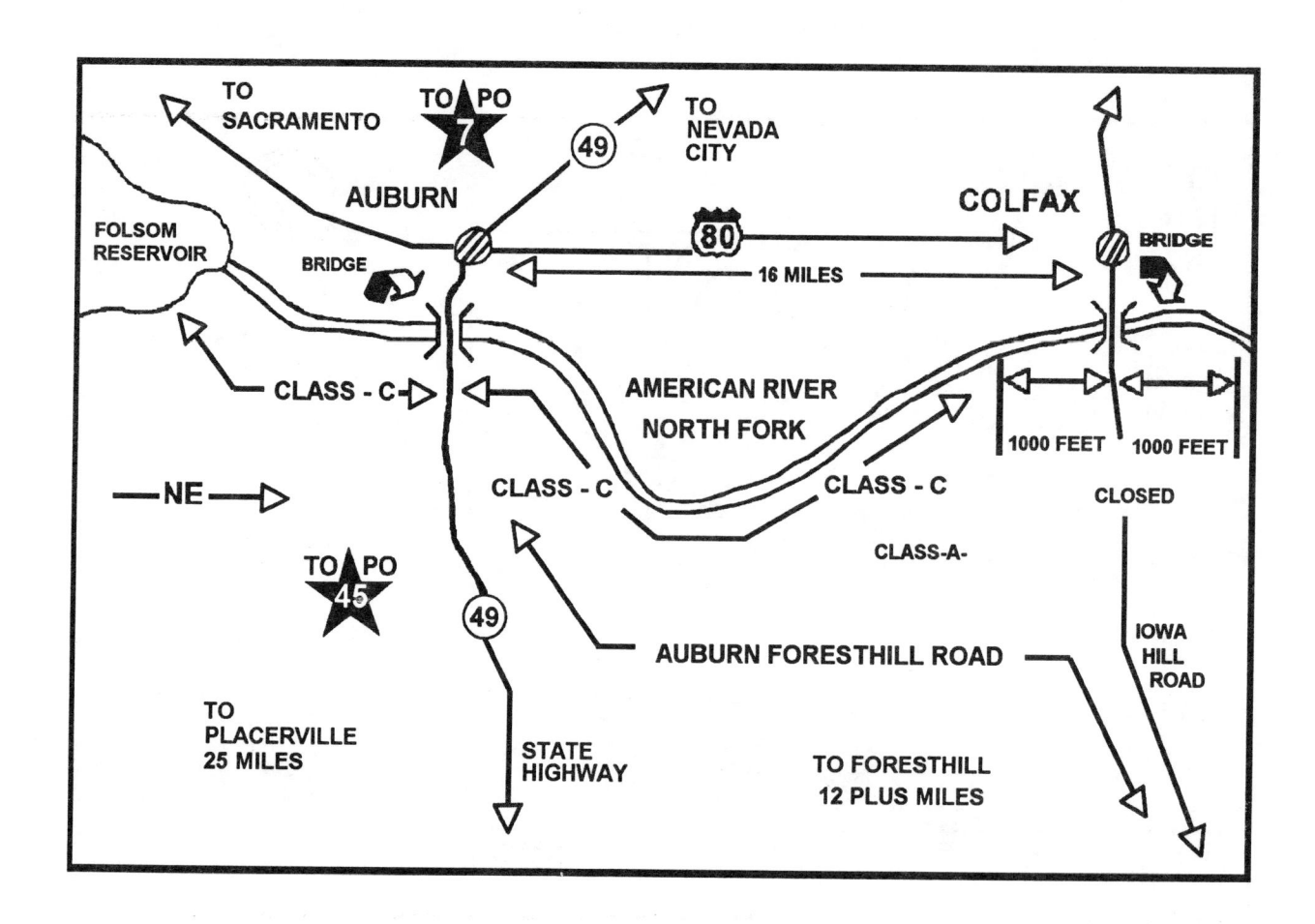

AMERICAN RIVER, NORTH FORK OF THE MIDDLE FORK, (PLACER COUNTY). From its junction with the Middle Fork of the American River upstream to the bridge on the Michigan Bluff (Deadwood) Last Chance Trail at location T15N R12E Sec. 32 is. **CLASS- H.** Use the El Dorado National Forestry Map, Tahoe National Forestry Map, California Highway Map and the Placer County Map for locations.

Adjacent to the present American River lay stream gravels and sand deposits of the Victor Formation, where most of the gold values remain within the terrace gravel along the banks. The greater portion of this area is underlain with granodiorite intrusion, with several northerly lenticular bodies of serpentine hosting rich gold-quartz veins with stringers fingering up onto the surface.

The Auburn State Recreation Area is one of the largest urban-accessible *OPEN TO THE PUBLIC AS A FREE-USE RECREATIONAL GOLD MINING SITE* in the State of California. Normally, vacuum-suction dredging is prohibited in any of the State Park units within California. Under a special exemption, recreational gold dredge mining, gold panning, gold sniping with a metal detector and gold searching within the back-country are permitted under certain conditions within this lucrative area of the Auburn State Recreation Park. (for info. write or call Calif. Parks & Rec. Auburn Rec. Area 7806 Folsom Auburn Rd., Folsom, CA 95630 916-988-0205.)

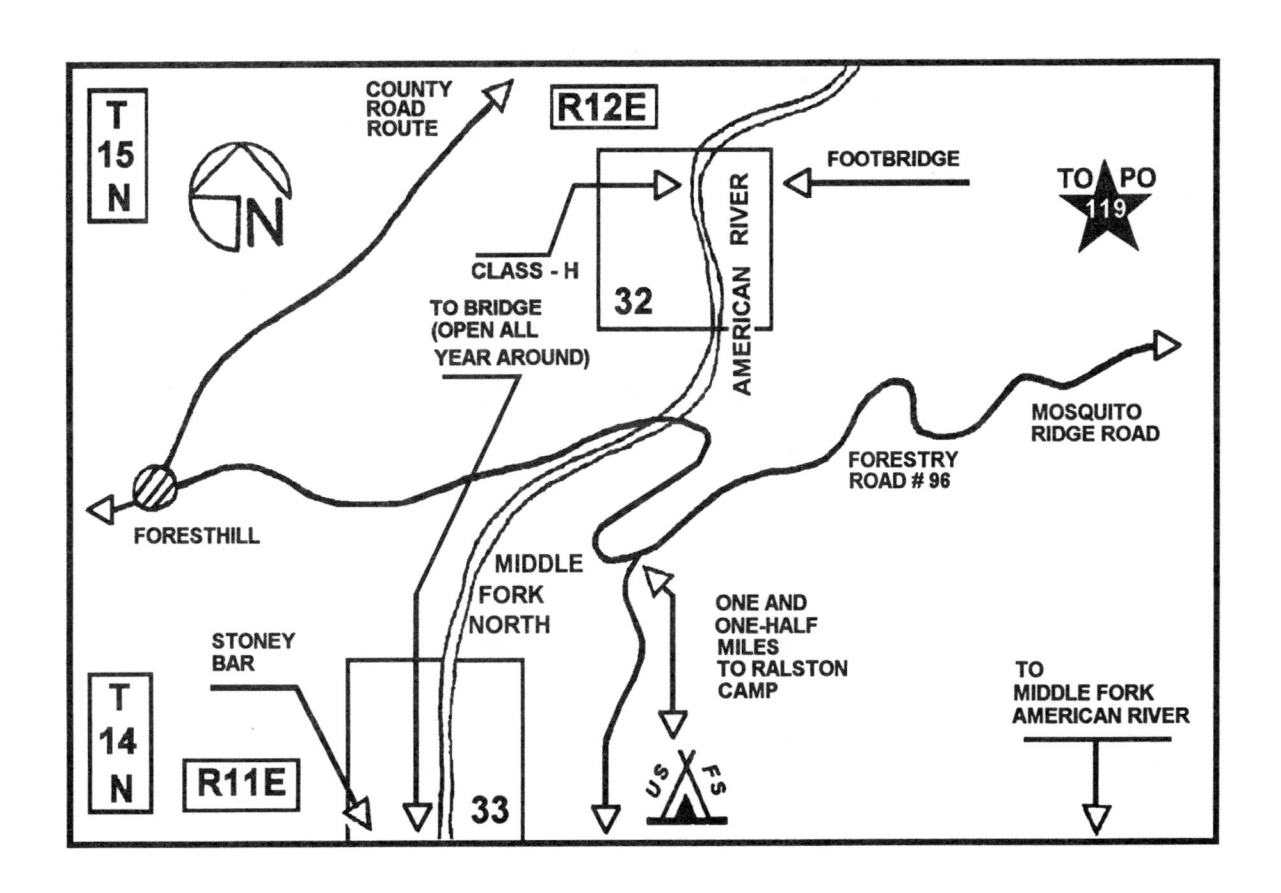

AMERICAN RIVER, MIDDLE FORK, (EL DORADO and PLACER COUNTIES). From its junction with the North Fork of the American River upstream to the confluence with the Rubicon River is *CLASS - C.* **OPEN TO THE PUBLIC AS A FREE-USE RECREATIONAL GOLD MINING SITE.** Filing of mining claims are prohibited. A California dredging permit is required when dredging in the water-runs. Gold panning is a free-use activity in California.

Dredges are limited to intake diameter of less than eight inches. Dredging is prohibited in the North and Middle Forks of the American River from Highway 49 bridge upstream to the Old Foresthill Road Bridge on the North Fork and to Louisiana Bar on the Middle Fork. Highbanking is prohibited and dredging material used for sluice boxes shall come from permanent running stream beds only. (Write or call for license permit info to; Calif. Fish 'n Game License 'n Revenue Branch, 3211 "S" St., Sacramento, CA 95816. 916-327-0195.)

The pre-Tertiary channel of the Middle Fork of the American River enters the area at Michigan Bluff and fingers out into the back-country area spreading its influence. The aged quartzite gravels are typical of the region near bedrock with its gold of a coarse nature and cemented firmly within its grasp. The greatest yields have come from gold quartz veins near the serpentine bedrock that have leached into the gravel of the creeks and water-runs in the adjacent area, which is the norm for the East and West Belt of the Mother-Lode Country.

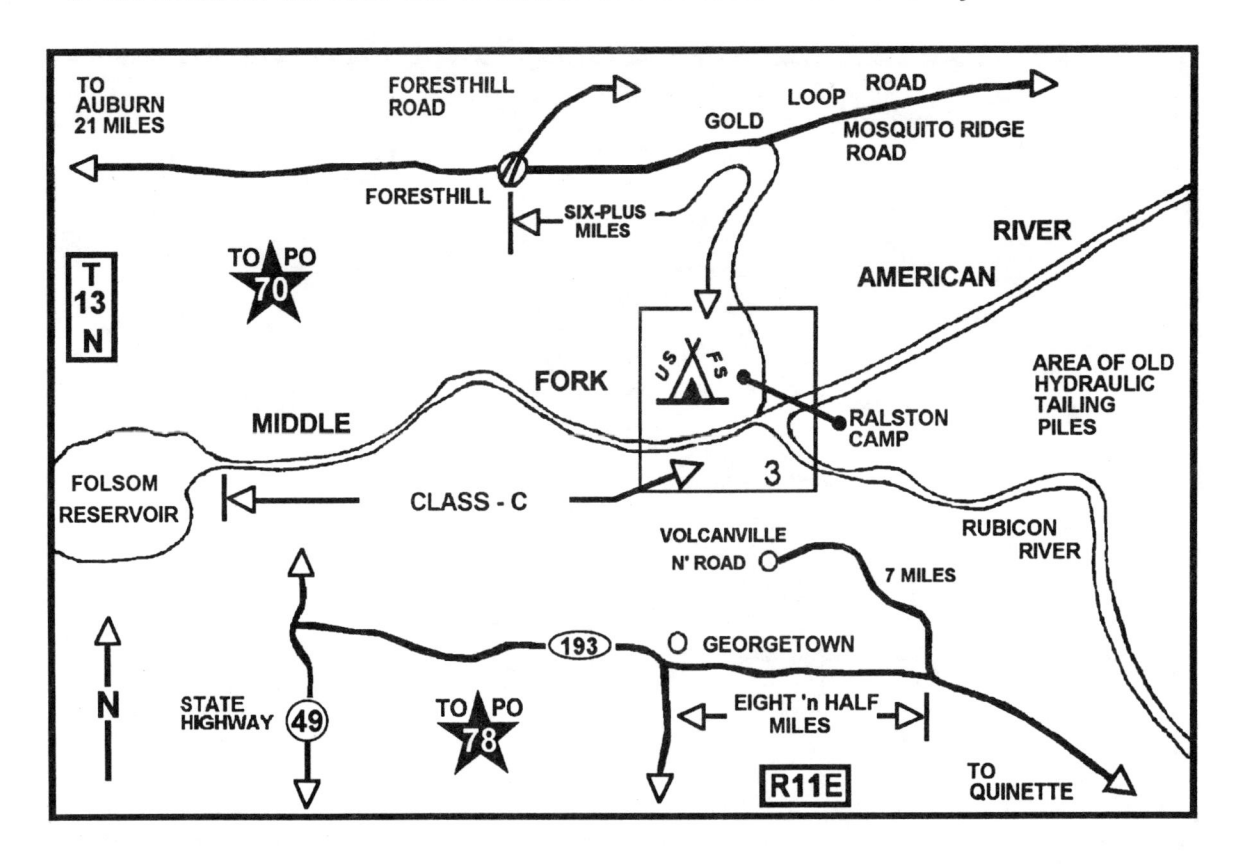

PLACER COUNTY. East of Highway # 49 is **CLASS - C.** Open to dredging from the fourth Saturday in May thru October 15. Remainder of the County is **CLASS - H.** Open all the year around except where noted. Use the Tahoe National Forestry Map, the El Dorado National Forestry Map and the County and California Highway Maps.

Considerable gossans of oxidized pyrites with rich quartz ore and abundant sulfides were found to be protruding from the gravel surface in the early days of mining. The possibility of these same conditions to be existing today, and with the raking from the thunderstorms tearing down the gold-bearing particles can be feasibly present in the back-country.

Most of the County is interlaced with ancient channels bearing extensive clear quartz gravel and placer deposits. Hydraulic tailings are in evidence from the early day mining activities and deserve careful searching of those piles for overlooked gold and platinum nuggets. The modern day dredgers working the water-runs near Foresthill find old quartz gravel close to the bedrock with well cemented gravel to be yielding the most gold.

Feeder creeks adjacent to the North Fork American River hold un-even bedrock formations consisting of deep and rich potholes of placer gravel with rich yielding benches and terraces; Indian Creek, Brushy Canyon, Shirtail Canyon and all of their feeder freshets have this distinction. Michigan Bluff area in the early days of mining produced an enormous amount of coarse gold. The area rivers, their tributaries and feeder creeks continue to tear from the gravel channel benches and Terraces of its free-gold deposits, sending the gold particles and gold occurrences into the lower riffle holding basins.

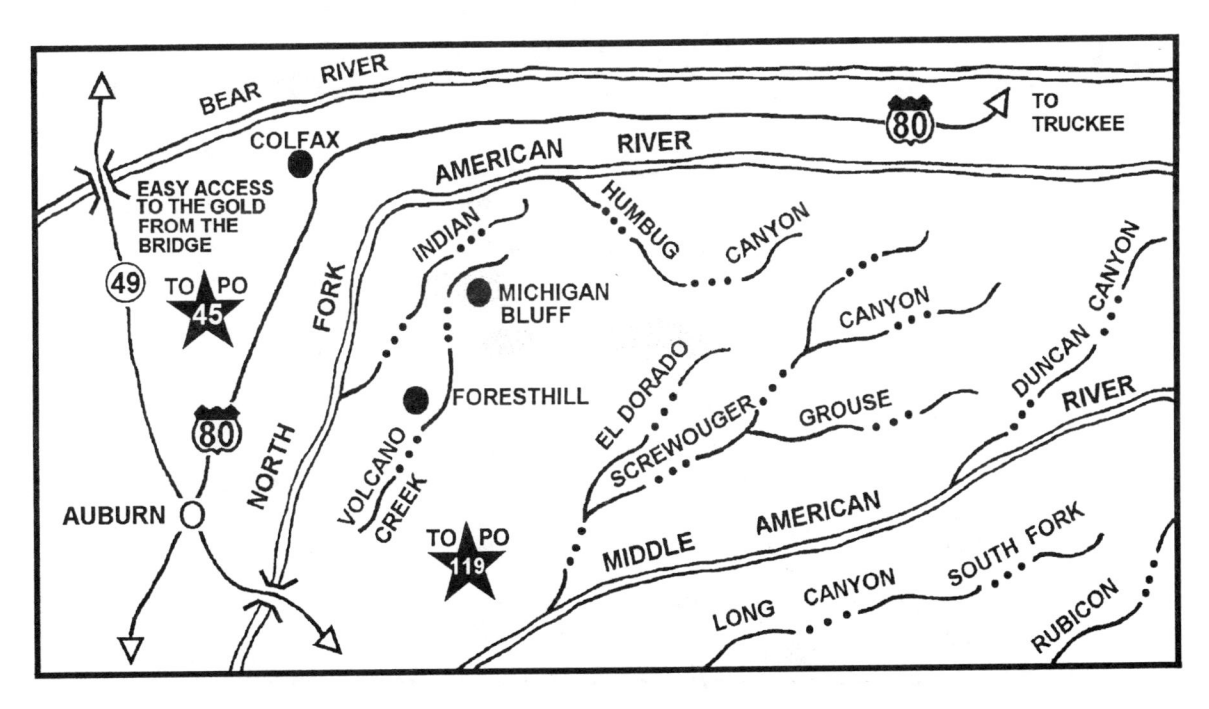

VOLCANO CREEK (PLACER COUNTY). From Mosquito Ridge Road located at T14N R11E Sec. 30, upstream to the Paragon Mine Tailings Dump, located at T14N R11E Sec. 30, is **CLASS H.** Open all year around except where noted.

PLACER COUNTY, east of Highway # 49 is **CLASS - C.** Open to dredging from the fourth Saturday in May thrue October 15, except where noted. The remainder of Placer County is **CLASS H .** Open all of the year except where noted. Use Tahoe National Forestry Map, County Map, California Highway Map and the Topographic Quadrangle Map for locations. Be alert to private property and existing mining claims. Water-runs of Placer County cut ancient channels carrying gold particles that eventually get dumped into re-newed riffles and holding basins of the back-country. Southwest and west of Foresthill flows the famous Volcano Creek that cuts thrue a series of intermittent small depressions of channel gravel from the pre-volcanic period.

Much of the ancient channel influence leads north from the Middle Fork American River and the Mosquito Ridge Road, and strings out into fingering spur channels past the Paragon Mine and the Mayflower Hydraulic Mine. Several miles of tailing piles show vivid evidence of work performed by these two majors. About 50% of gold particles have been left behind on the tailings due to poor mining methods of greed and haste. With each heavy rain-fall and water run-off raking the tailing heaps and moving those overlooked gold particles from the tailings into the riffles of the adjacent tributaries. Much of the walls and the banks of the ravines, gorges and the mini-canyons have gold bearing gravel making up the substance in the rhyolitic tuff sectors hosting the gold.

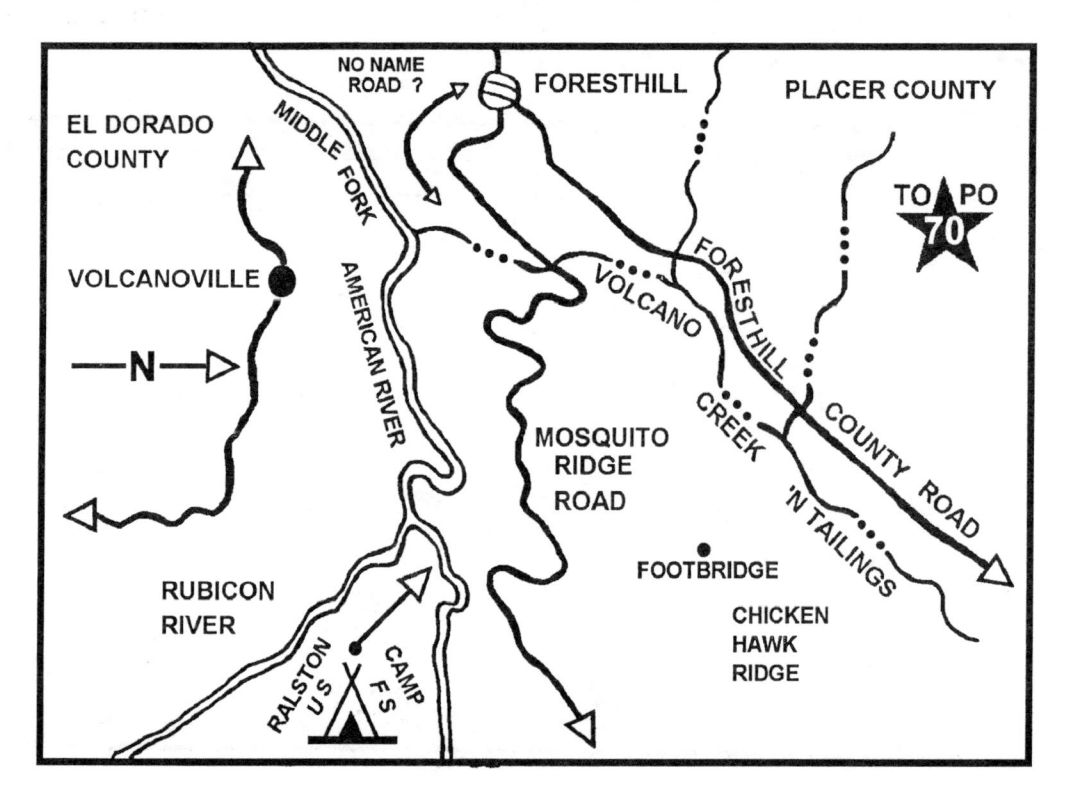

85

NEVADA COUNTY. East of Highway # 49 is **CLASS - C.** Open to dredging from the fourth Saturday in May thrue Oct. 15. Remainder of County is **CLASS - H.** Use the Tahoe National Forest Map, the County and California Highway Map for location. Be alert to the many mining claims and private property in this gold lucrative bearing district. Check with the County Recorder's Office for land status. A history of placer mining, hydraulic and dredge mining within the ancient lower gravels at the bedrock level contained mega-rich gold occurrences, large nuggets with an occasional diamond falling into the sluice box. The upper layer containing a medium value in scattered placer gold particles was influenced by seasonal washing from springtime thunderstorms.

Much of Bear River and its tributaries, the Wolf River and its feeder creeks, the Middle Fork Yuba River on the north County line and its many tributaries, and the South Yuba River flow thrue the middle of the County with its tributaries, were influenced by the southwest trending Tertiary channel of the ancient Yuba River that carried quartz-rich gold-bearing gravels. This heavy influence was dropped into narrow-intermittent basin pockets, previously created by pre-volcanic forces filling those cavities with rich placer gold.

Most roads of the area follow alongside of the creeks, or else the creeks will be crossed by these same roads at some point in its flow. The country-side holds many old mining camp sites that were frequented by people that lost items, gold coins and bank holes with buried pouches filled with gold nuggets. The search with a metal detector will uncover the caches.

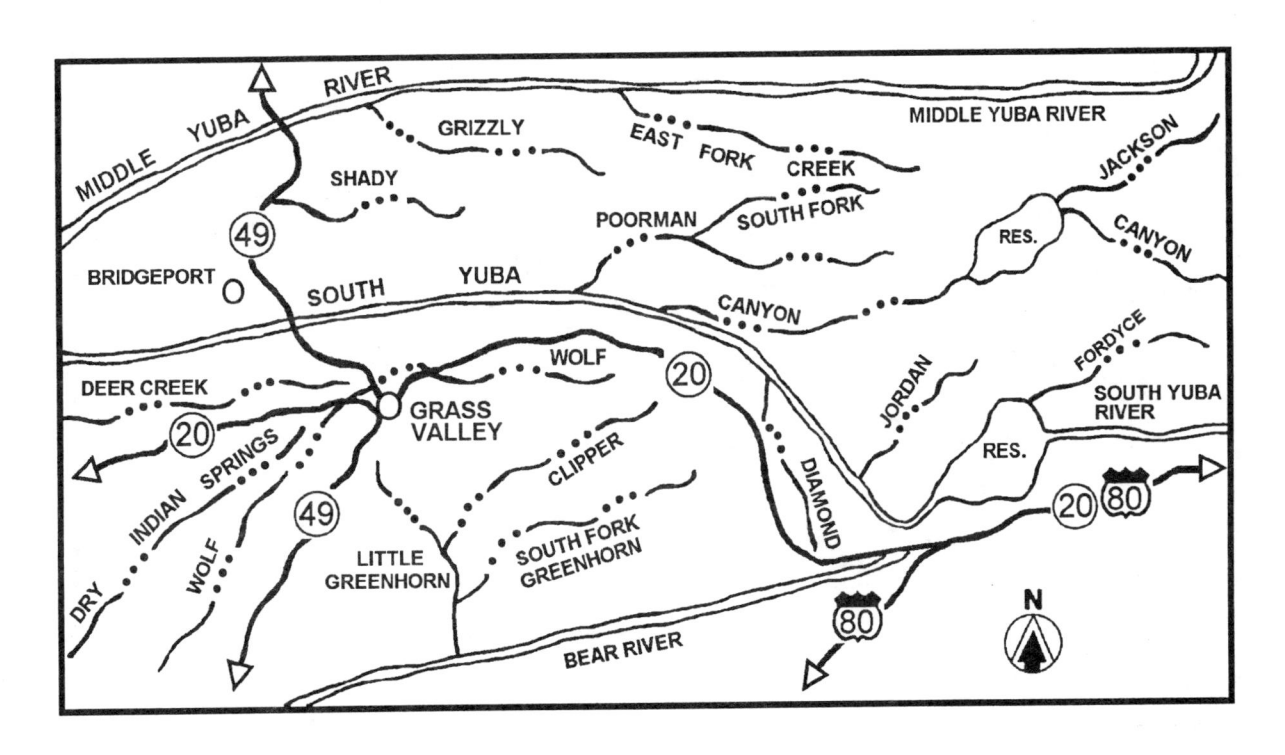

BEAR RIVER (NEVADA AND PLACER COUNTIES). From the bridge at Highway # 49 upstream to the Dutch Flat Powerhouse at location T16N R10E Sec. 27 is **CLASS - H.** From the Forty-Mile Road to the South Sutter Irrigation District's dam is **CLASS - D.** Use the National Forestry Map, Nevada and Placer County Maps and California Highway Maps for location.

From Auburn to Dutch Flat via I-80 Highway is about 28 miles, with feeder road access to Bear River off of I-80 and State Highway # 174. Be alert to existing mining claims and private property by determining land ownership thrue the Lands Office of the BLM, County, State and USFS records.

Bear River during heavy rain fall and spring run-off is the recipient of gold particles from the adjacent influences of the Mother Lode Gold Belt. There is an enormous accumulation of tailings on Bear River that holds an approximate amount of 50% of the gold placer that the old miners let slip thrue their inefficient method of mining. The seasonal erosion from the Neocene Yuba gravels (auriferous gravels) bring with it gold particles, platinum, diamonds, gold nuggets that are dumped into the surrounding feeder-creeks of the area, which in-turn releases its gravity burden into the gravel bars, riffles and catch basins of the Bear River.

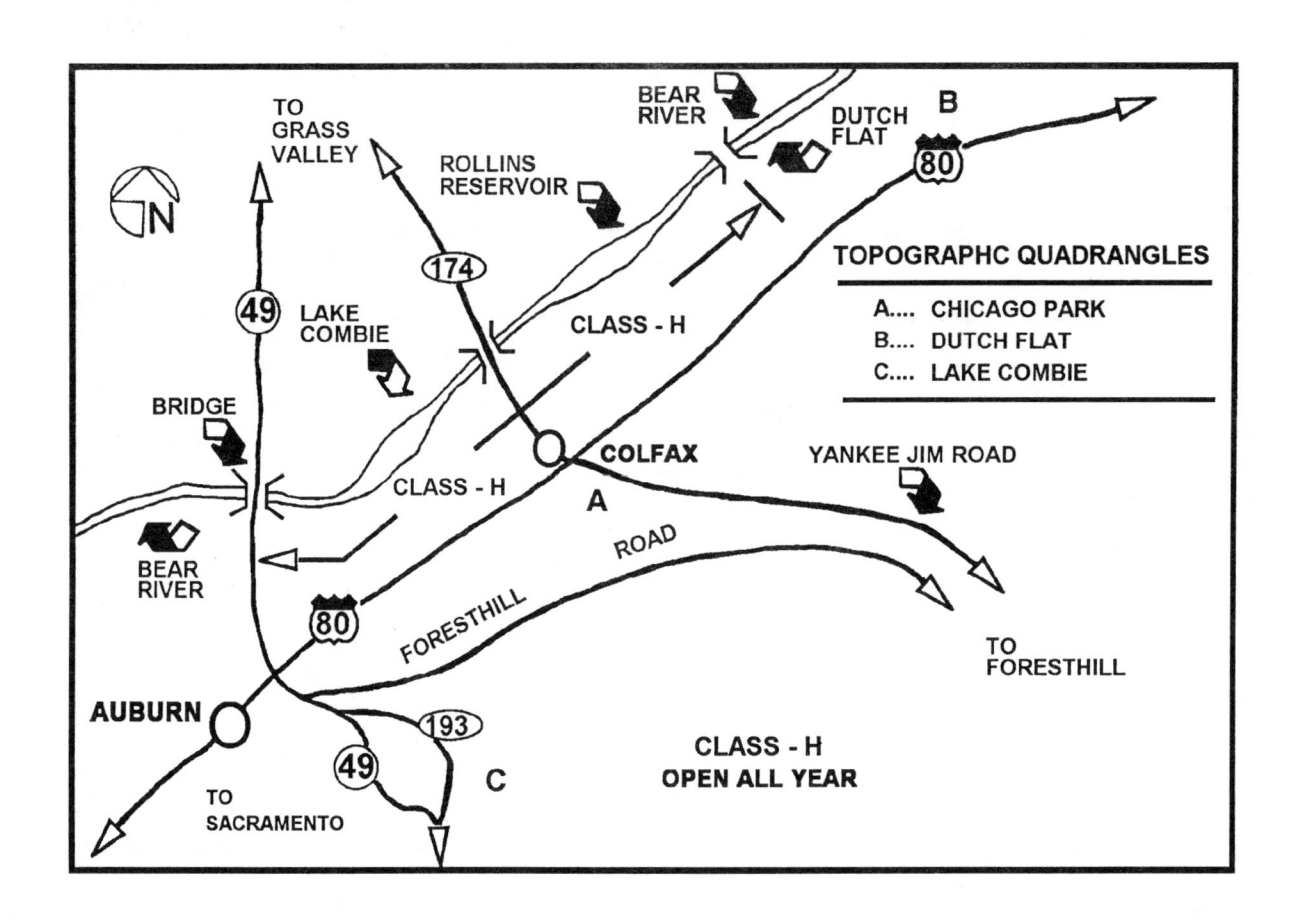

WOLF CREEK (NEVADA COUNTY). From the Tarr Ditch Diversion at the location of T15N R8E Sec.10, upstream is **CLASS - C.** Open to dredging from the fourth Saturday in May thru October 15 except where noted. Use the Topo Quadrangle *Wolf and Lake Combie* for location.

SHADY CREEK (NEVADA COUNTY). From the junction with the South Yuba River, upstream is **CLASS - H.** Use Topo Quadrangle *Nevada City* for exact location. **GREENHORN CREEK (NEVADA CREEK).** From the mouth at the Rollins Reservoir at T15N R9E Sec. 2, upstream to the Buckeye Road at location T16N R9E Sec.19 is **CLASS - H.** Open all year as noted. Use Topo Quadrangle *Chicago Peak* for location.

STEEPHOLLOW CREEK (NEVADA COUNTY). From the junction with the Bear River at T15N R10E Sec. 6, upstream to Camels Hump Rd. is **CLASS - H.** Open all year. Use Topo Quadrangle *Chicago Peak* for location. Check for existing mining claims with the County Recorders Office for accessibility rights in the area of interest. *For the identification of the above named creeks and surface road positions, use the Tahoe National Forestry Map, California Highway Map, County Map and the Thomas Street Atlas Road Map for Nevada County.*

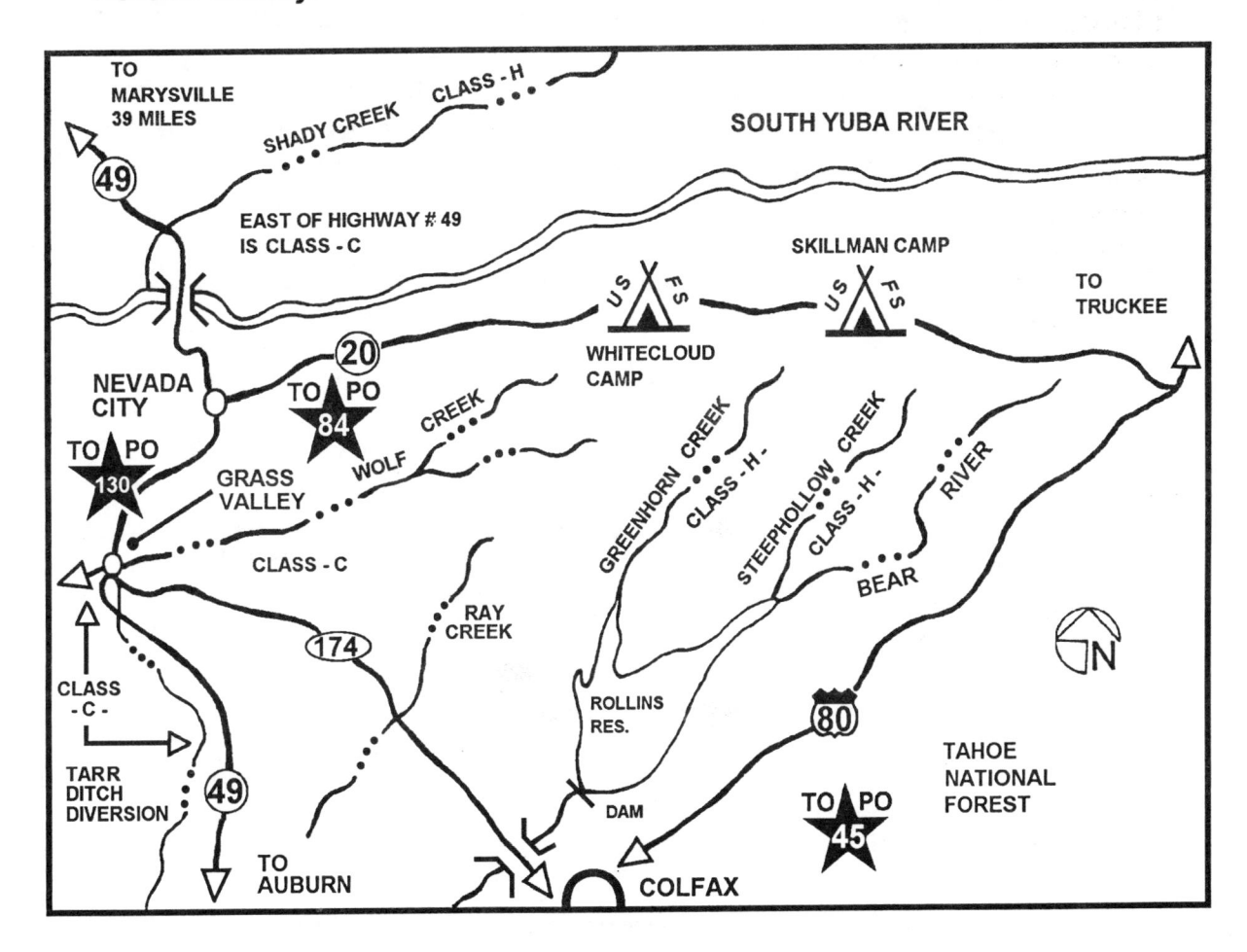

SISKIYOU COUNTY IS *CLASS - E.* Open to dredging from July 1 thrue September 30, except where noted. Use the Klamath National Forestry Map, County and California Highway Map for location. South-Central Siskiyou County experienced considerable mining from the old bench gravels, canyons and gulches that had dropped the placer gold particles into the upper Scott River region at location T40N R8W. Bucket-line dredges operated in this district, disturbing and re-locating gold-bearing gravel that with today's modern metal detector can seek out those overlooked gold particles where the old miners had failed to claim. Just north of Fort Jones, at location T43N R9W, important water-runs were heavily worked thrue the late 1800's. The area continues to be worked by gold snipers and recreational miners in Cherry Creek, Deadwood Creek, Indian Creek, French Creek, McAdam Creek and all of their tributaries and feeder creeks.

Happy Camp at location T16N R6W is a prime location for establishing a base camp in preparation for entering the back-country to do some work on the tributaries of the area. Water-runs flow thrue dozens of rich, but dominating gravel deposits that had been established by pre-Cretaceous older stream channels with a succession of terraces and benches that range several yards above the present flowing channel and its water-runs. At Happy Camp, the Klamath River struggles around several sharp bends and effects the gold particle drop and settling area. The gravel benches of the area saw much hydraulic mining, where the inefficient dredging methods deposited half of the retrieved gold back onto the tailing piles and sluiced debris dumps. Be alert to the gold content of the piles 'n heap debris by hand raking the cobbles and passing a metal detector over each raking stroke. Klamath River with feeder creeks near Happy Camp area are; Indian Creek, Clear Creek, Buzzard Creek, Oak Flat, East Fork Indian, Five-mile and Ten-mile creeks and Thomas Creek. Be alert to mining claims and private property.

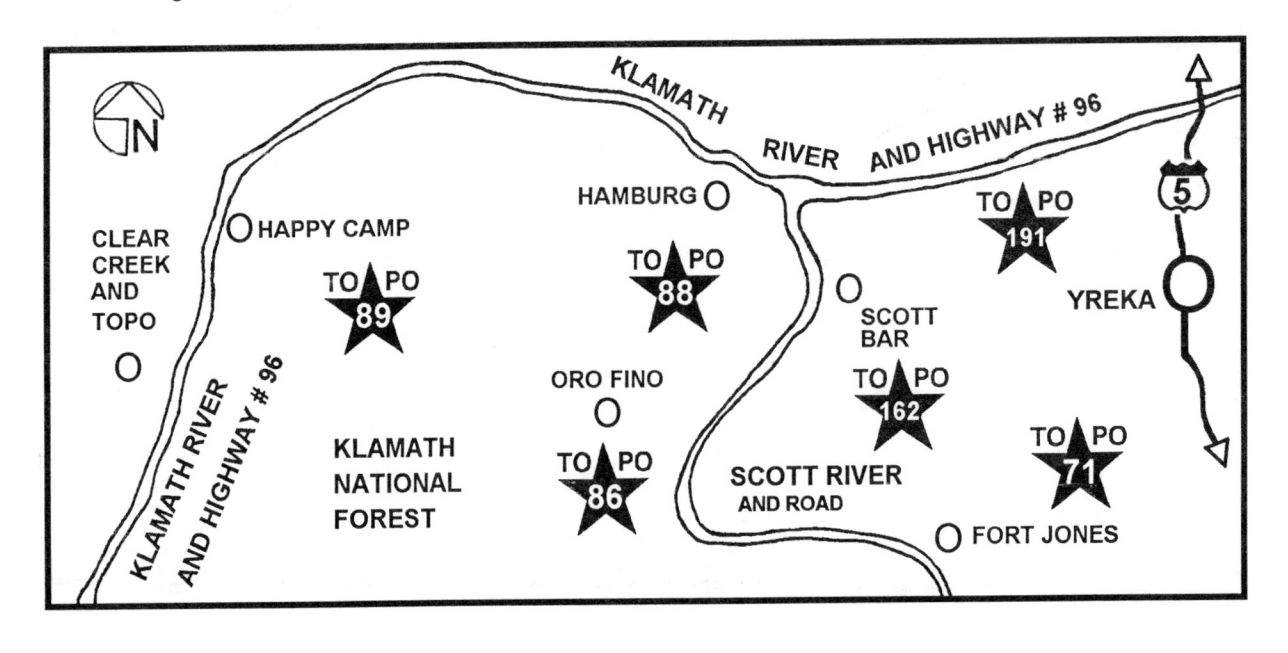

SCOTT RIVER (SISKIYOU COUNTY). CLASS - G. Open to dredging from the fourth Saturday in May thru September 30. Use Topo Quadrangles, *Callahan, Mc Conaughy Gulch, Fort Jones, Scott Bar and Hamburg for locations.* Scott River junctions with the South Scott River and the East Fork Scott River at location T40N R8W Sec.17. Scott River ends at the confluence of the Klamath River at location T45N R10W Sec.6. Use Siskiyou County Map, California Highway and the Klamath National Forestry Map. Older bench gravels exist along both sides of the Scott River and finger out from the adjacent gulches and their tributaries to the river. The bench gravels were extensively mined by bucket drag line dredging, hydraulic and ground sluicing. Check those tailing piles and heaps for those overlooked gold nuggets left behind by the old time miners.

Cherry, Indian, McAdam, Deadwood and French Creeks have yielded enormous amounts of placer gold in the past, with Cherry Creek having been worked six times over, (which means that the back-country is releasing the gold occurrences from time to time into the Cherry Creek, even for today). Currently, part-time gold searchers, dredgers and snipers are seasonally active in locating those **gold particles** from the gravel benches that the watery run-off drop into the catch-all basins of Cherry Creek. The area is humungus in size with many tributaries hosting older gold-bearing Tertiary bench gravels with its gold being distributed thrue-out the gravel strata. Gold particles not necessarily being locked only into conventional bedrock, but also are to be found existing within the cemented hardpan of the false bedrock.

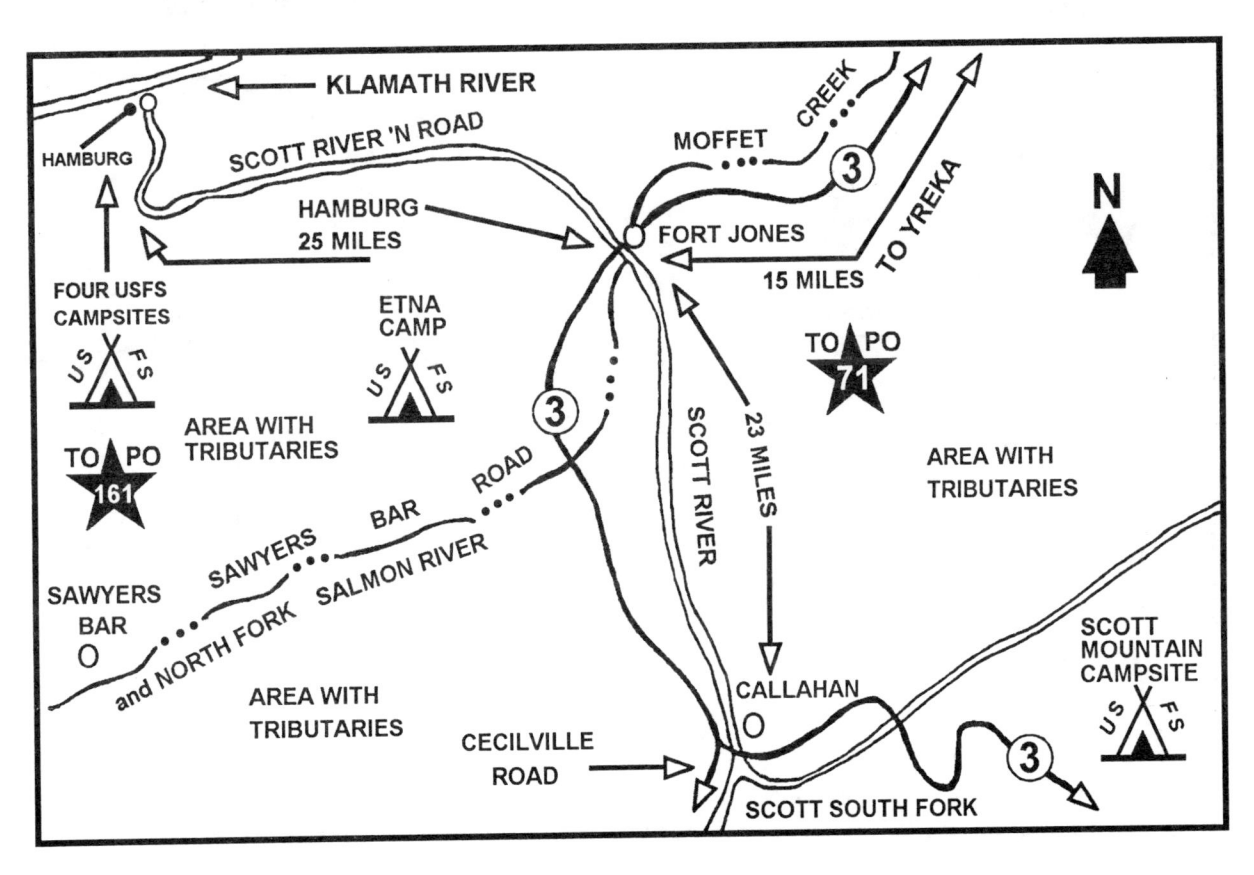

BOGUS CREEK (SISKIYOU COUNTY). CLASS - E. Open to dredging from July 1 thrue September 30. Use Siskiyou County Map, California Highway Map and Klamath National Forestry Map; location T47N R5W Sec.16. Use Topo Quadrangle, *Copco & Hornbrook.* Leave Henley-Hornbrook at US I - 5 and travel south, then east on Copco Road approximate 4-miles to junction with Ager Road. Turn right on Ager Road and travel southeast for about 2.8-miles to junction with Ager-Beswick Road. Turn left onto Ager-Beswick Road and travel northeast for about 6 miles to bridge crossing at Bogus Creek location at T47N R5W Sec. 13.

To the east of Hornbrook are tributaries at location T46N/T47N R4E that drain from the Klamath National Forest and the dominate landscape figures of Eagle Rock and Black Rock Mountain Peak Ranges; with Bullhead Creek, Cold Creek, Parker Creek. Be alert to mining claims and private property. Check with the County Recorders Office for accessibility rights in the area of interest.

History of gold production in and around Hornbrook at the confluence of Bogus Creek, Cottonwood Creek and their tributaries where the area was noted for extremely rich shallow deposits. Gold is found in gravel of present stream channels that come from ancient terrace and bench deposits adjacent to those water-runs......Check those tailing piles 'n heaps with a metal detector for overlooked gold and platinum nuggets. The old-time mine labors failed to check the long toms with sufficient care by allowing the overlooked gold particles to pass over the riffles and slip out the end of the flume onto the tailing piles.

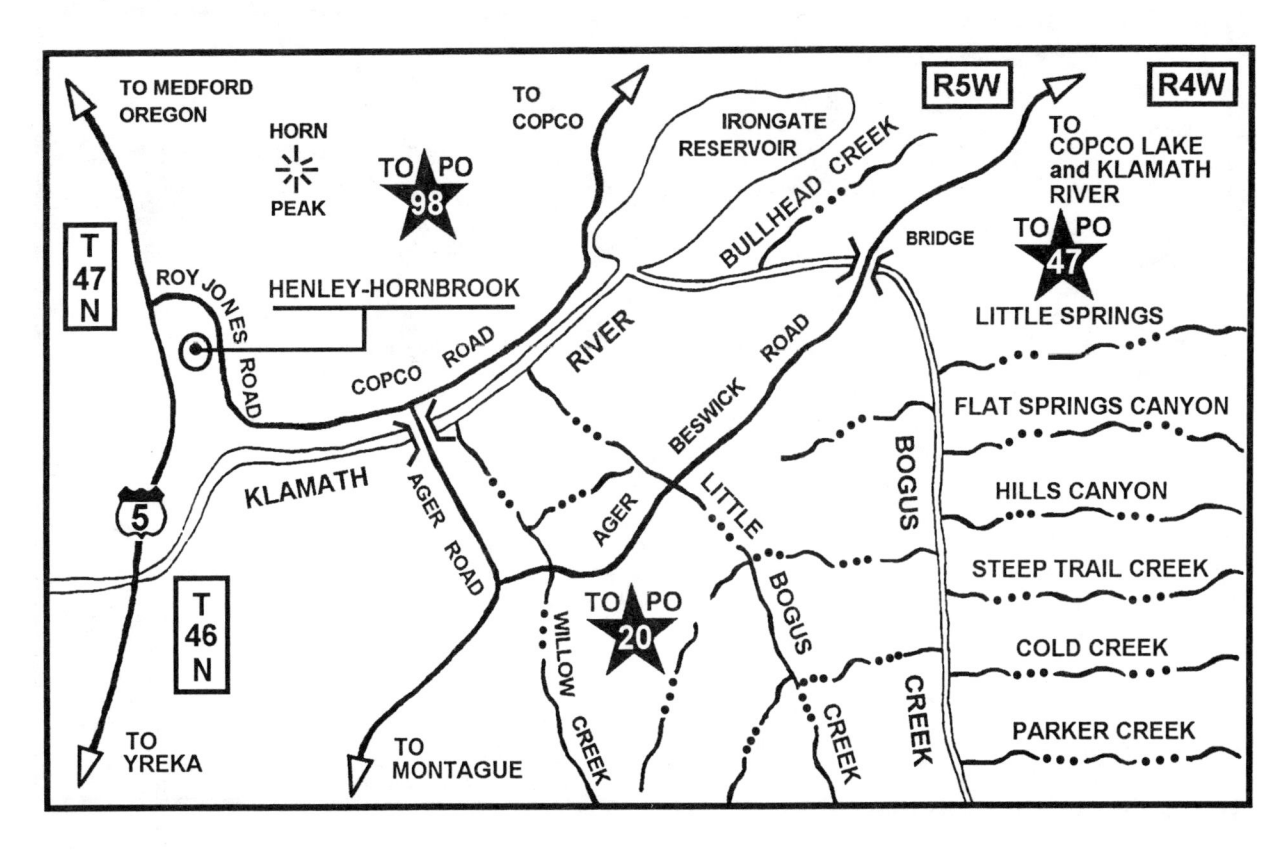

GREENHORN CREEK (SISKIYOU COUNTY). Greenhorn Creek above the City of Yreka Reservoir is *CLASS - H.* Open all year around. Use Siskiyou County Map, and at location on the Klamath National Forestry Map is T45N R1E Sec. 29. Between the years of 1850 and 1900 the Greenhorn Creek produced eleven million dollars in **placer gold**. The area had been extensively placer-mined of extremely rich deposits during the early years of dredging and hydraulicing mining.

Present gold particles come from the surrounding tributaries that rake the adjacent older gravel banks, terraced channels and ancient holding basins. The **back-country** hosts exposed ledges of quartz veins containing gold pyrite and smaller degrees of **sulfides.** Seasonal rains leach and percolate those native gold particles and gold occurrences into the area's water-runs and their holding basins.

Tributaries of the area are responsible for replenishing the gold content of the area water-runs; Mill Creek, Soap Creek, Copper Creek Lime Gulch, Cherry Creek, Timber Creek, Punch Creek and the Scott Bar Mountains. Humbug Creek to the north has experienced considerable drag-line activity thrue the early years of mining. Check those tailing piles and debris heaps at the mouth of those drift mines where the miners had missed by inches the rich gold-bearing deposits of the pre-Cretaceous gravel channels. Be alert to mining claims and private property. Do research work at the County Recorders Office for the area of interest.

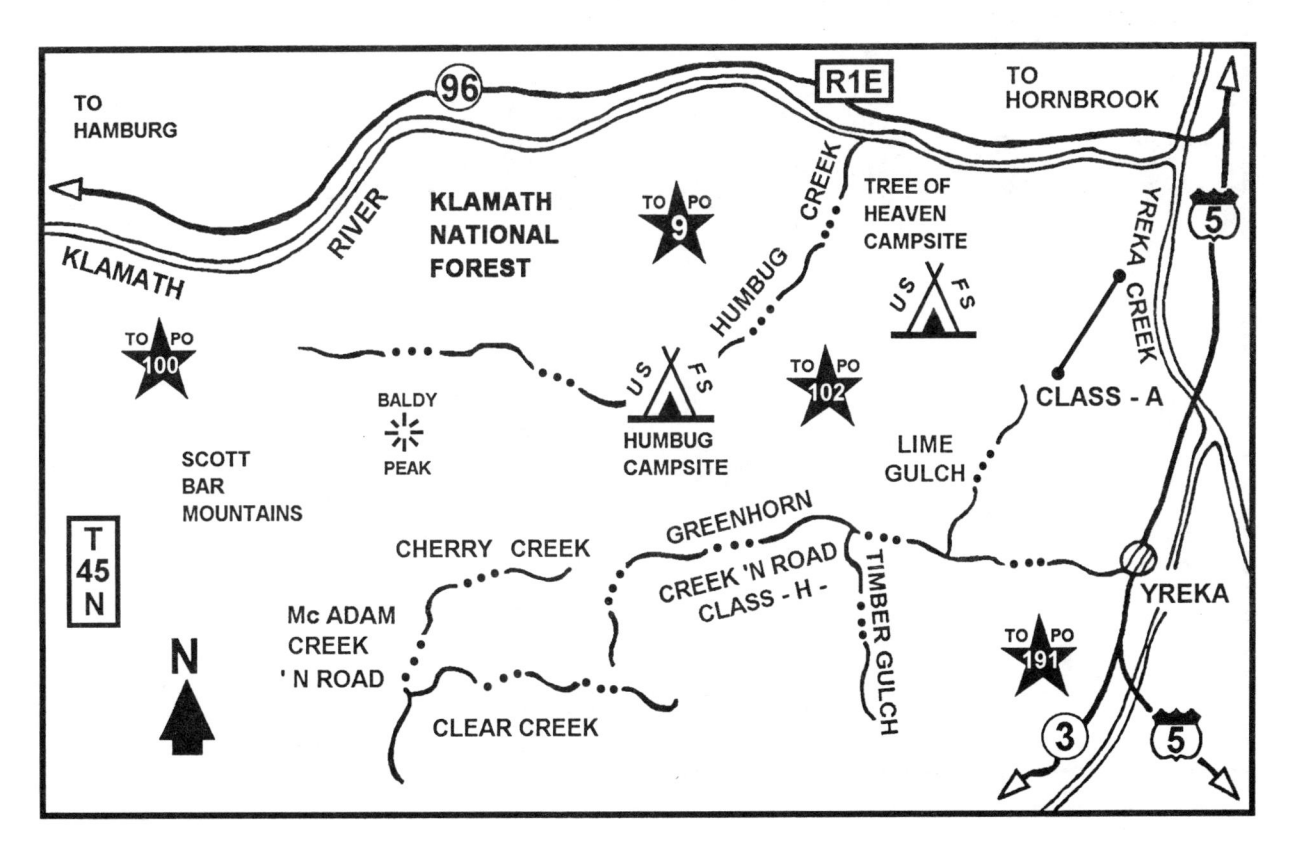

SHASTA RIVER (SISKIYOU COUNTY) CLASS - A. Dredging allowed by special dredge permit only. Use Topographic Quadrangle for location *Hawkinsville, Montague, Lake Shastina and Weed.* Use Siskiyou County Map, Klamath National Forestry Map, California Highway Map and Shasta Trinity National Forestry Map.

Tributaries of Shasta River are; Dry Gulch, Julien Creek, Willow Creek, West Forks Parks Creek, Eddy Creek, Dale Creek, Boles Creek, Beaughton Creek, Little Shasta Creek. Most of the tributaries have been fed from the gulches on either side of those water-runs carrying **ancient gold placer** from the pre-Cretaceous gravel deposits. Look for ridges that have been untouched and overlooked that carry pockets of detritus material that has eroded from the edges of the faulted portions of a large channel that is equivalent to the old Tertiary run. The Old Blue Shore gravel extends beneath the Shasta Valley influencing all of the tributaries with its placer gold in segregated small quantity spotty pay gravels. Extending north of Granada within the Shasta Valley is a spur stringer from a remnant of a ridge that once dominated the area with its ancient shore line gravels. The ancient gulches running off from the ridge carried gold and platinum into the channels that formed the future drainage system of the present Shasta Valley. Gravels that haven't eroded away and continue to reside in their original spotty position where they host the pre-Cretaceous gold 'n platinum particles, remain undiscovered in the back-country.

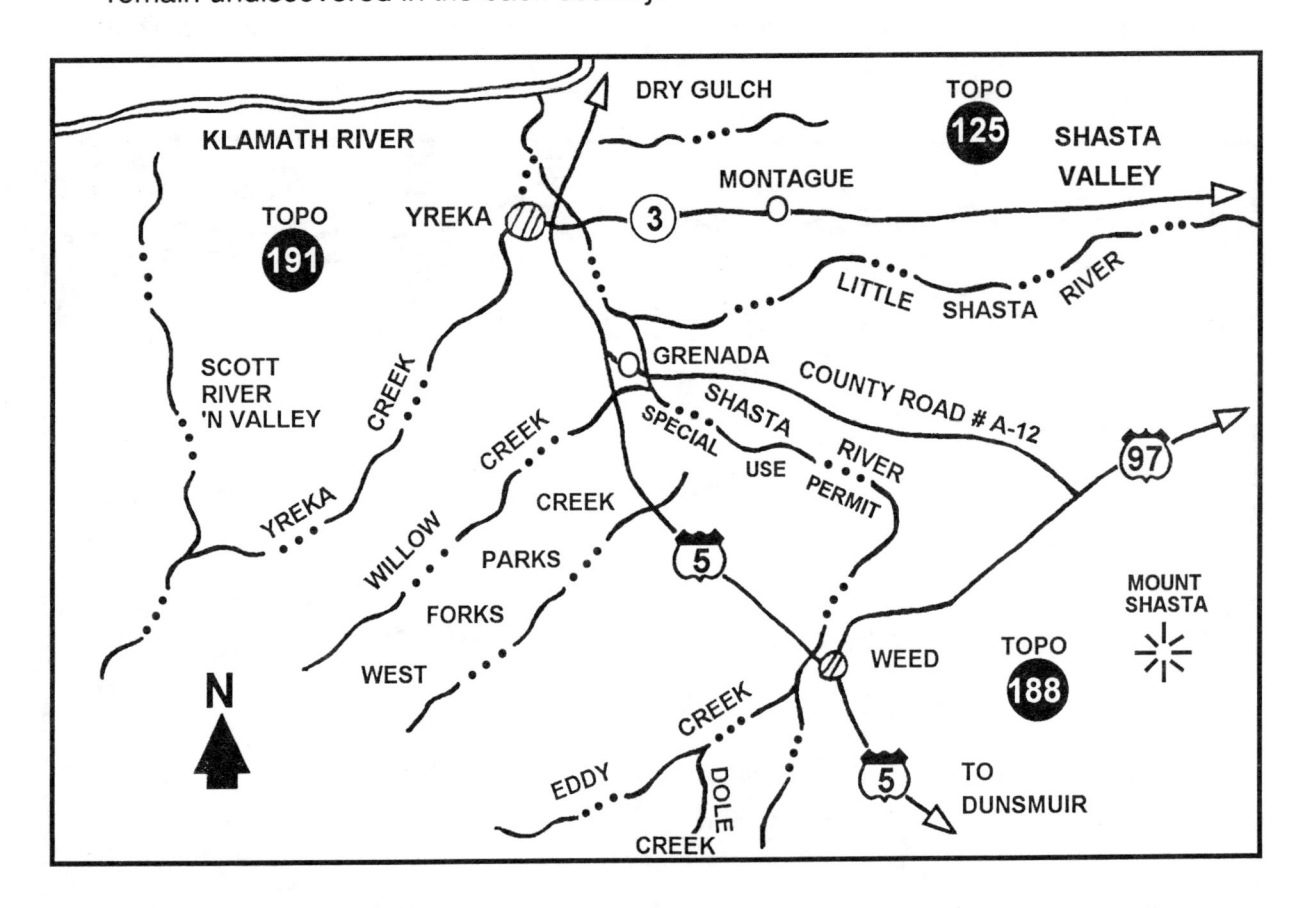

KLAMATH RIVER (DEL NORTE, HUMBOLDT and SISKIYOU COUNTY).

From the mouth upstream to the Salmon River is **CLASS - G,** from the Salmon River upstream to 500 feet downstream of the Scott River is **CLASS - H,** from 500 feet downstream of the Scott River upstream to Iron Gate Dam is **CLASS - G,** and from Iron Gate Dam to the Oregon border is **CLASS - A.**

KLAMATH RIVER TRIBUTARIES and their Tributaries, except the main stem TRINITY RIVER (DEL NORTE, HUMBOLDT, SISKIYOU and TRINITY COUNTIES); where dredges with intake nozzles larger than six-inches are prohibited.

Klamath River flows over the northern sector where old benches, terraces and ancient channels with gravel lay in position and are seasonally cut by present flowing water-runs that rake loose much of the Late Jurassic gold into present day holding basins. Suction dredgers, sluicing and gold snipers are influenced into working those holding basins of the adjacent tributaries where disturbed **gold particles** have been deposited.

Much of the Klamath River flows around sharp bends depositing its noble metals into drop points much like what is experienced at Happy Camp. Check out those old tailing piles 'n heaps with a metal detector and unearth those overlooked gold nuggets.

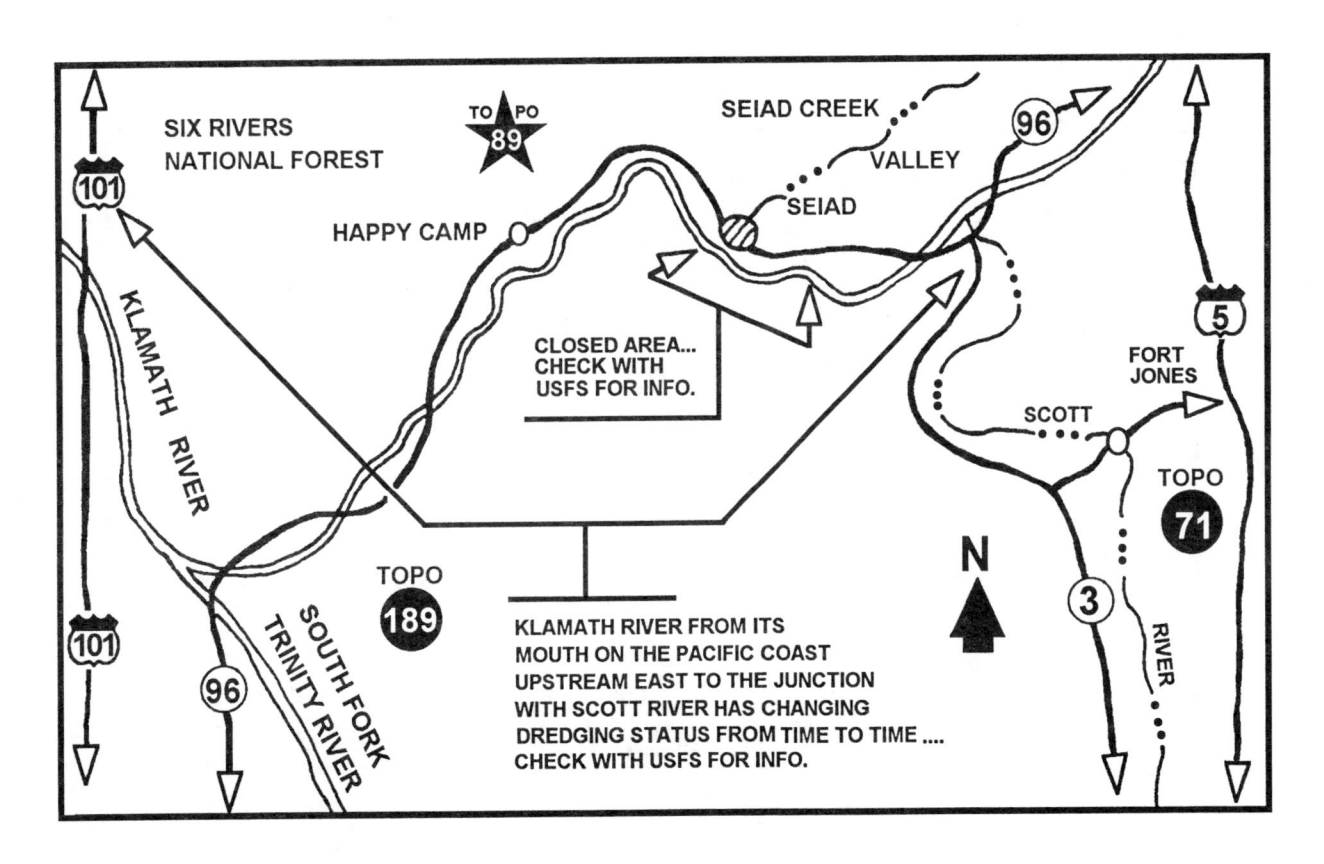

HUMBOLDT COUNTY. *CLASS - E.* Open to dredging from July 1 thrue County and the California Highway Maps for location.

In the past the Klamath and Trinity Rivers with their tributaries and feeder creeks gave up much of the gold. Numerous older Humboldt County terraces and bench gravel deposits that lay above the present river levels and those contingent to modern day water-runs, are raked by each seasonal rainstorm sending its percolated ancient gold particles into the lower creeks for settlement. The ancient gold that is located is fine to medium with platinum present in the magnetite black sands.

Along the Klamath and Trinity River complex system the old time hydraulic miners used fine screens in the trommel, and the result, heavy gold passed on thrue and out onto the piling tails. Hydraulic operators working the many old channel beds worked at a clumsy, fast and sloppy pace to get at the gold. By digging with a careful, slow and methodical pace and by putting everything thrue the sluice, all of the bedrock gold would have been claimed.

The old miners with their incompetent fast-pace mining operation failed to use a larger screen on the staker dredgers. Because of this missing mining component, well over 50% of the ancient gold being processed was dumped back onto the staker piles. Should there be debris heaps laying in a haphazard position, this will indicate to the gold searcher that this land mass was carelessly mined and will be a "must" area to be gold sniped with a metal detector, and with the possibility for a one-man rocker-sluice operation. From the County line along the Klamath River to the junction of the two rivers, south past Hoopa on the Trinity River, mining of any sort is prohibited on Indian lands.

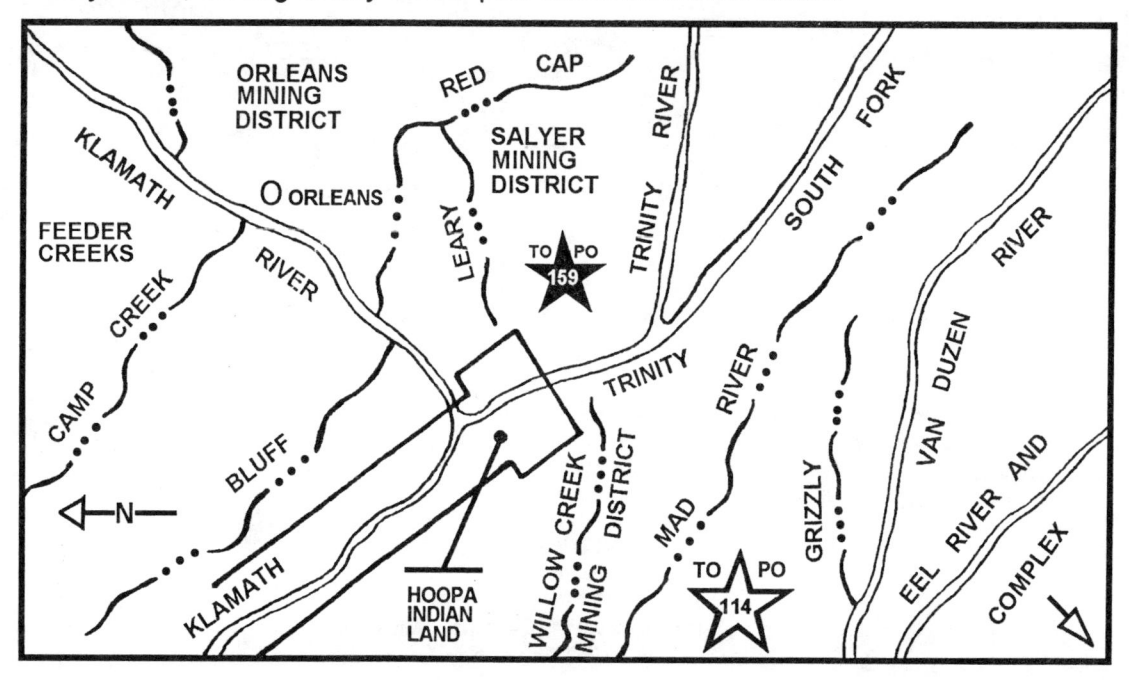

YUBA COUNTY....*CLASS - H.* Open to dredging throughout the year except where noted. Use the Plumas National Forestry Map, County and California Highway Maps for location. The area of Camptonville, T18N R8E location, hosts old hydraulic tailings that once held rich gold deposits that had been laid down by the old Tertiary Yuba River, along with several medium sized deposits of Tertiary Channel which contributed to the **gold** placer content. The area continues to release gold particles from the old Tertiary gravels into the surrounding water-runs. The location area T20N R8E once was a major placer field which was supported by the old Tertiary Channel deposits. Hydraulicing of the area left considerable amounts of **gold** particles remaining on the tailing heaps due to inefficient mining methods. Local **nugget** snipers and gold particle scavengers continue to search the debris heaps for overlooked **placer gold nuggets.**

The Hammonton District T16N R5E location is a major dredge field along an eight mile stretch of the lower Yuba River. The area was so rich with placer gold-bearing gravel that it supported twenty-one working dredges for several years of operation. The gold-bearing gravel of the placer field was estimated to be in the multi-million of cubic yards in gravel reserve, but are beyond the existing equipment capabilities for gold recovery at such deep depths.
Present mining has been curtailed.

Many of the water-runs flow thrue the "Blue Gravel" locations and are supplied with gold particles from the ancient Yuba River Channel and its many fingering Tertiary tributaries. With each thunderstorm that rakes the gravel surface creates a fluid percolation sending the loosened gold particles into the riffles of the present lower tributaries and their feeder creeks.

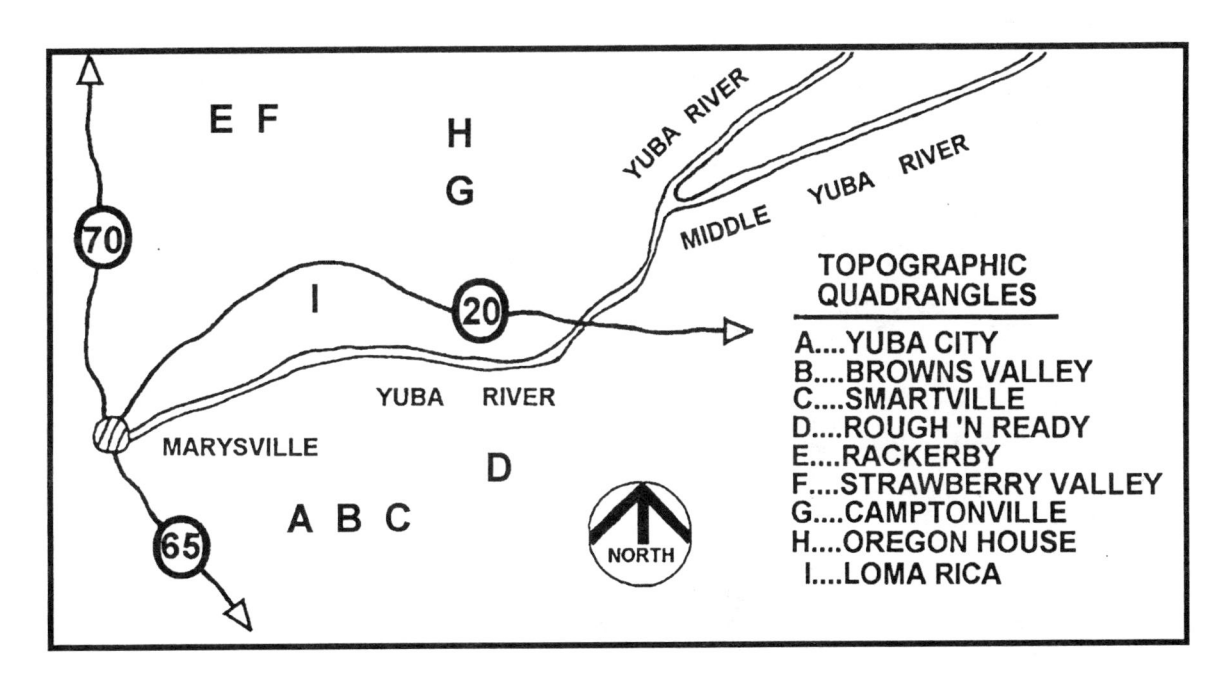

TOPOGRAPHIC QUADRANGLES

A....YUBA CITY
B....BROWNS VALLEY
C....SMARTVILLE
D....ROUGH 'N READY
E....RACKERBY
F....STRAWBERRY VALLEY
G....CAMPTONVILLE
H....OREGON HOUSE
I....LOMA RICA

WOODS CREEK BRIDGE at State Highway # 108 and # 49. TOPO; *Sonora.*
Location at T1N R14E. Use the Stanislaus National Forest Map, Tuolumne County Map, California Highway Map and the Automobile Club of Southern California Map. **A FREE-USE RECREATIONAL GOLD MINING SITE** located less than one-eighth mile on either side of State Highway # 108 and # 49 bridge that spans Woods Creek a couple of miles south of Jamestown.

This is a good place for getting used to newly purchased equipment while at the same time acquiring gold particles from one's working efforts. Located in, and influenced by the very heart of the Mother Lode Blue-Gray gravels of the ancient **deeply seated Tertiary channels.**

At any bridge which spans a creek has a Free-Use public domain right-away of sixty feet at the road proper. Mineral entry has been prohibited, allowing free-use with gold panning, dredging and gold sniping as a recreational gold mining activity. Good gold content shows up after each heavy rain run-off. This area receives special attention from the local dredgers who know that Woods Creek is influenced by the adjacent Tertiary placer gravels. Tuolumne Table Mountain lays near the north belt of Jamestown with Tertiary gravel deposits underlying the Latite formation in a SW-erly direction. Table Mountain was formed eons ago as a massive ancient drainage system which over the period of time was exposed to violent earthen upheavals and was re-positioned as a reverse, upside down river formation. Woods Creek eroded away Table Mountain by exposing its Tertiary gravel deposits of gold placer to continuous thunder storms. This area is rich with pockets of **placer gold** hidden from sight by their deep locations. Seasonal rains erode away the exposed gravel benches into the adjacent feeder creeks of the back-country, making this area a top choice for gold searching. Be alert to private property and existing mining claims.
Stay within the confines of Woods Creek bridge site.

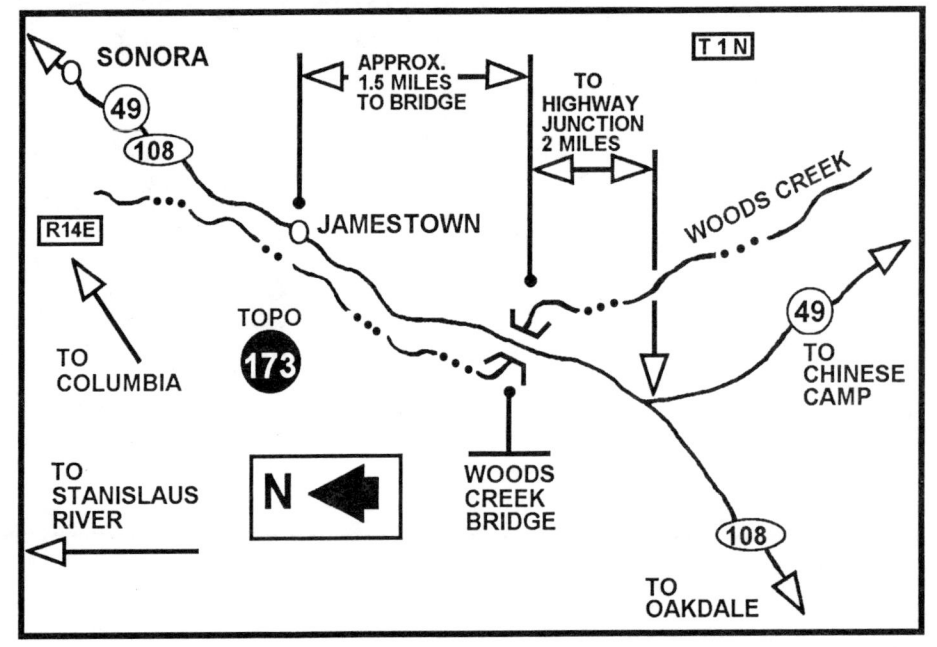

TUOLUMNE COUNTY, east of Highway # 49 is *CLASS - C.* Open to dredging from the fourth Saturday in May thru October 15, remainder of County is *CLASS - H* except where noted. For location, use the Stanislaus National Forestry Map.

WOODS CREEK AND TRIBUTARIES (Tuolumne County), location T1N R14E Sec. 33. TOPO; *Sonora and Chinese Camp.* Woods Creek and its tributaries from Harvard Mine Road (Jamestown) downstream are *CLASS - C,* from Harvard Mine Road upstream is *CLASS - A.*

KANAKA CREEK (Tuolumne County), location T1S R15E Sec. 7. TOPO; *Moccasin.*
CLASS - H, (open all year). Dredges with an intake larger than 4-inches is prohibited.

DEER CREEK (Tuolumne County), location T1S R15E Sec. 2. TOPO; *Groveland. CLASS - H,* (open all year). A dredge with an intake larger than four inches is prohibited.

BIG JACKASS CREEK (Tuolumne County), location T2S R15E Sec. 12. TOPO; *Groveland. CLASS - H,* (open all year). A dredge with an intake larger than four-inches is prohibited.

LITTLE JACKASS CREEK (Tuolumne County), location T2S R16E Sec. 5. TOPO; *Groveland. CLASS - H,* (open all year). A dredge with an intake larger than four-inches is prohibited.

CLAVEY RIVER (Tuolumne County), location T1S R17E Sec. 9. TOPO QUADRANGLE; *Jawbone Ridge. CLASS - A.*

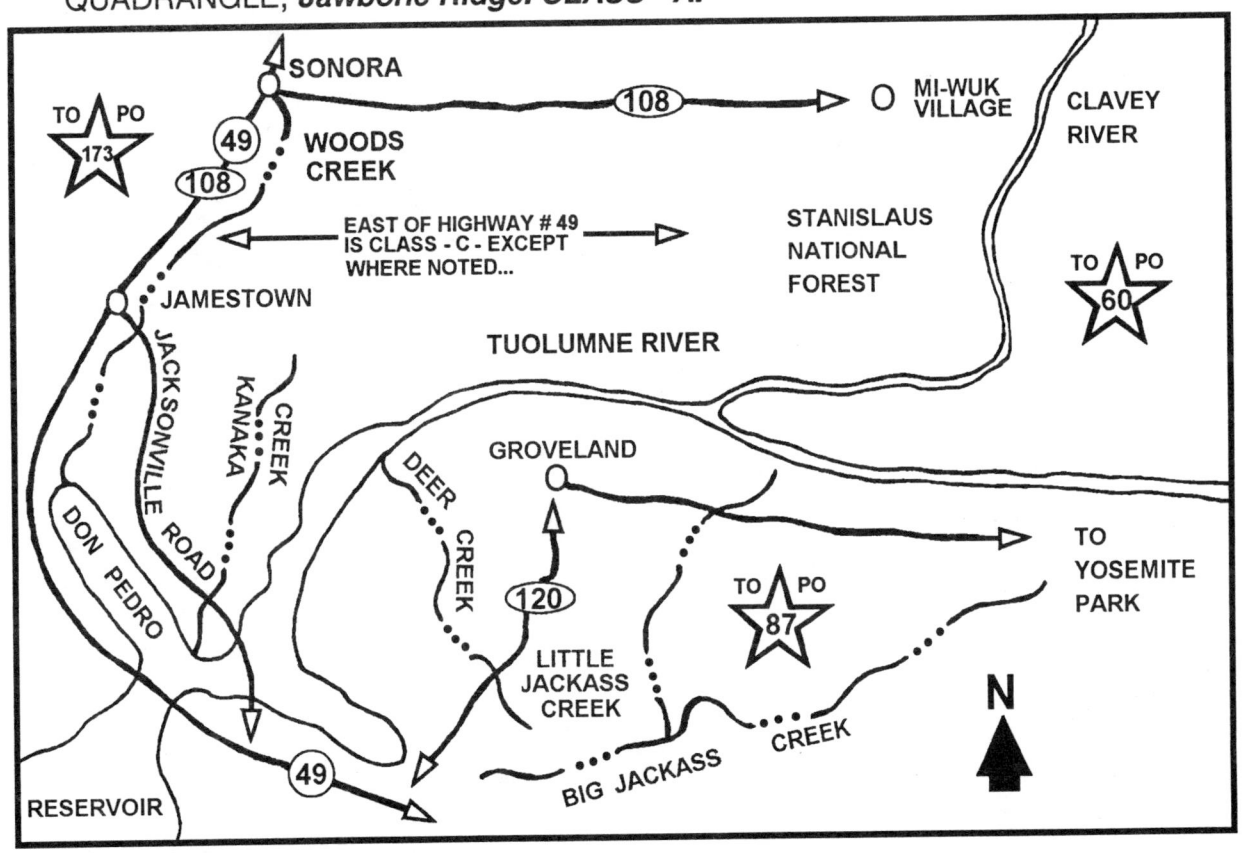

BUTTE COUNTY....IS CLASS C. Open to dredging from the fourth Saturday in May thrue October 15, except where noted. Use the Plumas National Forestry Map, California Highway Map and the Butte County Map for location.

Several isolated steep ancient channels finger their influence across from the north to the northeast down thrue the center of Butte County, skirting the eastern edge and exiting to the south into neighboring Yuba County.

The lower section of the Chico formation host gravel that contain gold particles of low to medium grade. With incessant winter rains, seasonal flooding and run-off eroding the gold particles from the confinement of the ancient gravel into concentrated patch basins making this one of the areas of interest. All of the water-runs, tributaries and feeder creeks that run their course thrue the ancient channels, Tertiary gravel, isolated basins and terraces are from the shore gravel of the Pleistocene Age and acted as carriers of placer gold, platinum, diamonds and exotic metals.

The isolated shore gravel and placer fields wherever they were located, produced rich amounts of gold particles for the old miners. It is believed that similar **placer fields** exist and have as yet to be discovered. Such gravel bodies lay deeper with an intermittent roller coaster formation influenced by ancient earthquake forces, giving adjacent water-runs their seasonal gold supplement and re-furbishing from the affect of heavy winter and spring thunderstorm erosion. Be alert to existing mining claims and private property. Check with the County Recorders Office for land status in the area of interest.

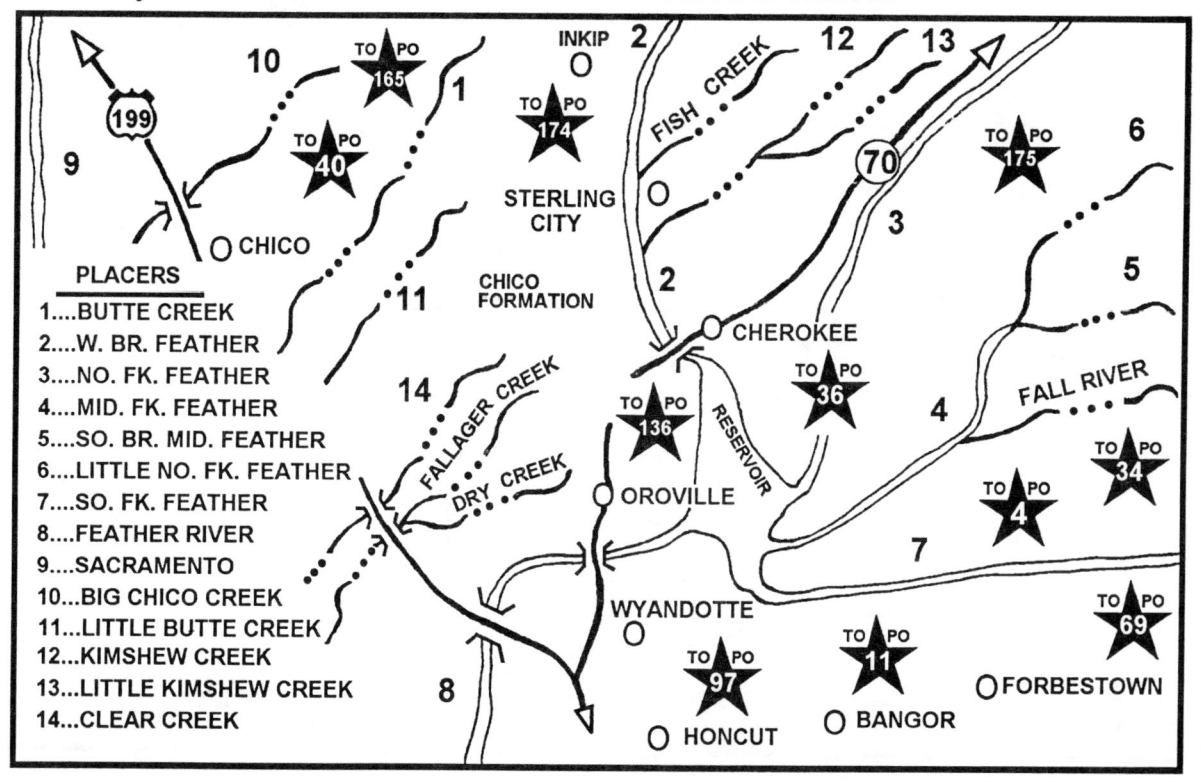

PLACERS
1....BUTTE CREEK
2....W. BR. FEATHER
3....NO. FK. FEATHER
4....MID. FK. FEATHER
5....SO. BR. MID. FEATHER
6....LITTLE NO. FK. FEATHER
7....SO. FK. FEATHER
8....FEATHER RIVER
9....SACRAMENTO
10...BIG CHICO CREEK
11...LITTLE BUTTE CREEK
12...KIMSHEW CREEK
13...LITTLE KIMSHEW CREEK
14...CLEAR CREEK

BIG CHICO CREEK (BUTTE COUNTY). From Manzanita Ave in Chico to the head of Higgens Hole, at location T24N R3E Sec. 31 is **CLASS - A,** closed waters with no dredging permitted at any time. Beyond Higgens Hole is **CLASS - C.** Open to dredging from the fourth Saturday in May thru October 15. Use Topo; *Cohasset, Paradise West, and Chico* Quadrangles for location.

MUD CREEK (BUTTE COUNTY). From its junction with Big Chico Creek upstream is **CLASS - C.** Use Topo; *Nord, Cohasset and Richards Springs for location.*

ROCK CREEK (BUTTE COUNTY). From its junction with Big Chico Creek upstream to the Butte-Tehama County Line is **CLASS - C.** Use Topo: *Nord, Richards Springs and Cohasset for location.*

BUTTE CREEK (BUTTE COUNTY). From the Sutter County Line upstream to the Durham Oroville Highway Bridge is **CLASS - H.** Open all year around. From the Durham-Oroville Highway Bridge upstream to the intake of Centerville Ditch at location T23N R3E Sec. 10 is **CLASS - A.** Closed waters. No dredging permitted at any time. From the Centerville Ditch intake at location T23N R3E Sec. 10 upstream, no dredge with an intake larger than four inches will be permitted.

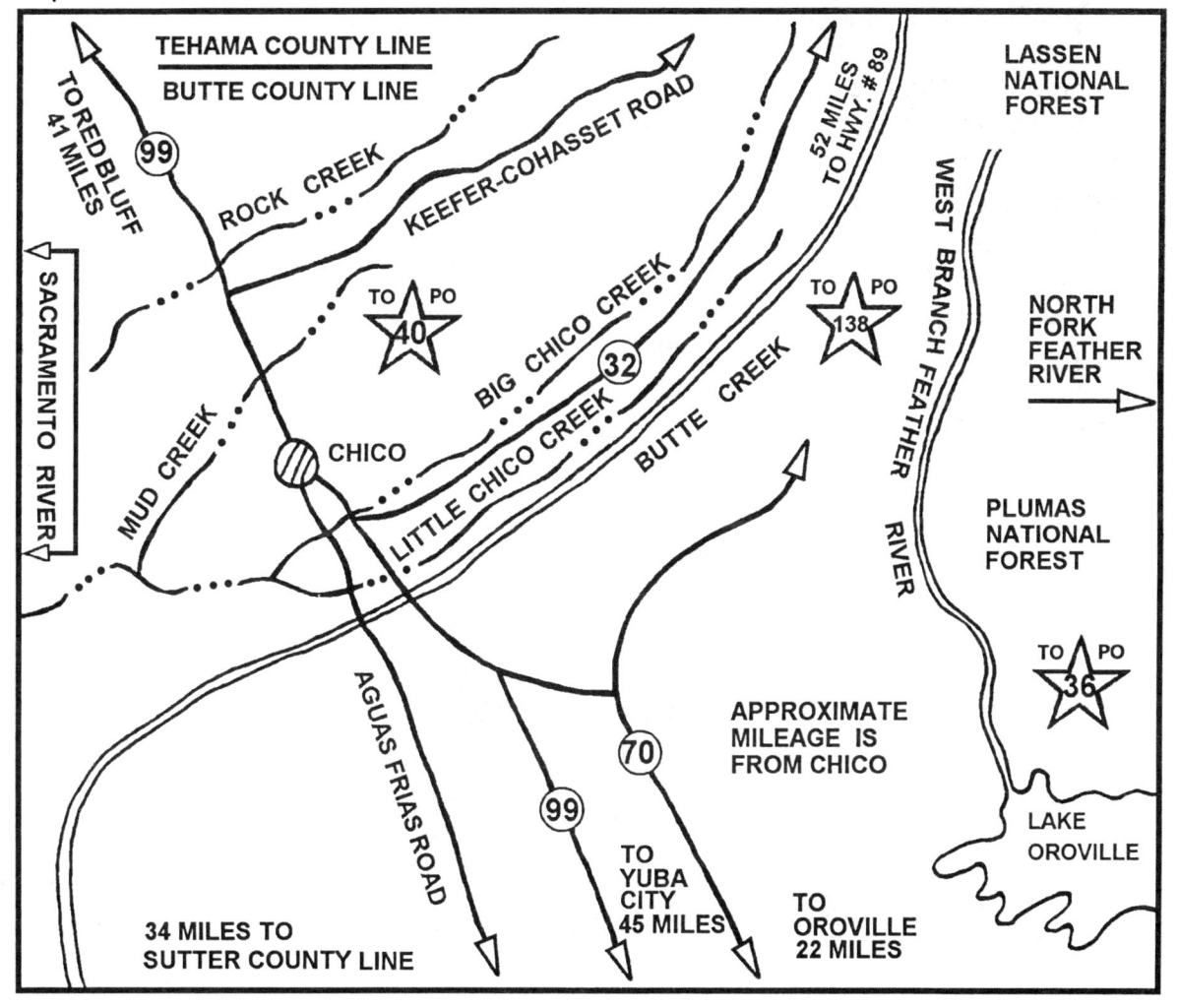

BUTTE CREEK MINERAL WITHDRAWAL GOLD MINING RECREATIONAL SITE. USGS TOPOGRAPHIC QUADRANGLES; *Cohasset and Sterling for location.* ***T23N-T24N R3E.*** **Filing of mineral claims are prohibited.**

In the north-central section of Butte County just above Paradise and northeast of Chico, sets an area provided by the Bureau Of Land Management (BLM) with thirty claim sites. For a small fee, permits can be obtained from; BLM, Redding Resource Area, 355 Hemstead Dr., Redding, CA., 96002 916-246-5325.

The Butte Corridor is open to the Public on a **first-come basis**. Exit Paradise and take Skyway north to De Sable. Go north and take the left fork to Ponderosa Way into the Butte Creek Canyon Way. For location use a Plumas National Forestry Map.

Ancient river channels flowed thrue this area when the country was a super-enormous island. Those channels hosted massive water-runs that carried tumultuous land debris with them that was thrashed around in vertiginous up-swells scattering the auriferous gold gravel to settle in the ancient riffles of those turbulent ancient water courses.

During the Cretaceous Period, Platonic forces moved those ancient river channels into haphazard locations exposing a wide and heavily mineralized porphyry dike containing levels of terraced stringers, basins and pockets filled with **gold nuggets** rough in size. Today's seasonal torrents of rain 'n flood run-off, rake the gold-bearing Tertiary gravel of the **back-country** and depositing its **golden charge** into those riffles held within the BUTTE CREEK CHASM.
A California dredging permit will be required should one engage in vacuum-dredging. **Good gold nugget-particle content.**

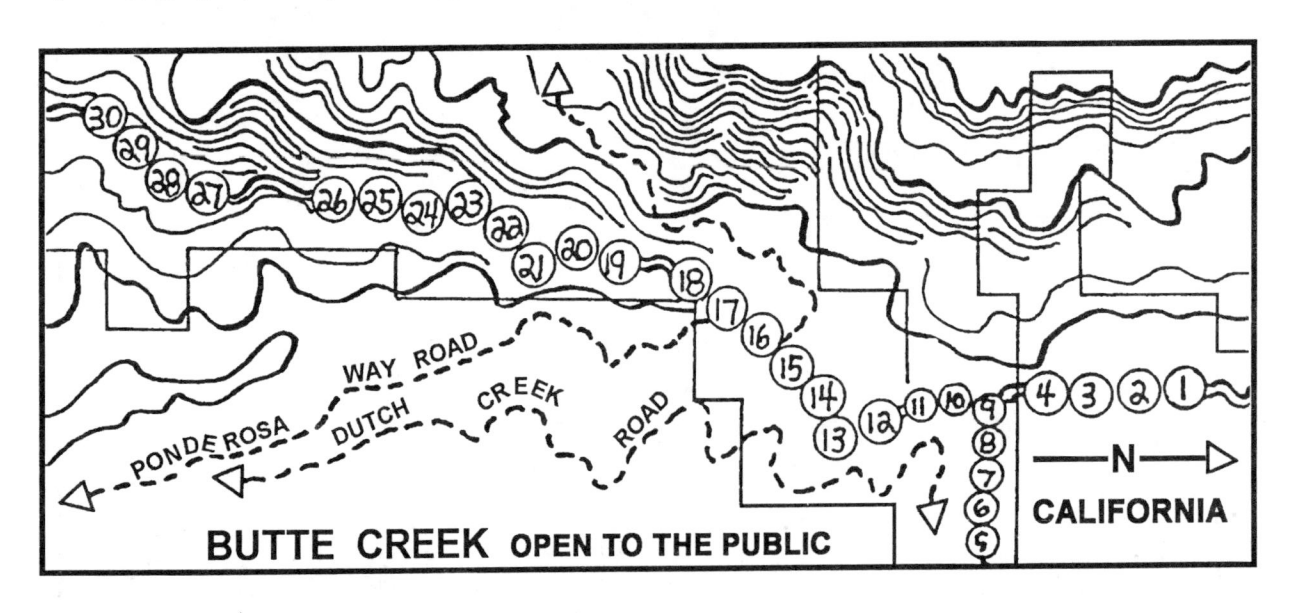

SACRAMENTO RIVER and TRIBUTARIES (several counties). The main stem of the Sacramento River from the San Francisco Bay upstream to Shasta dam is *Class - A.* The Sacramento River and its tributaries from Shasta Lake upstream to Box Canyon Dam are *Class - A.*

FEATHER RIVER (BUTTE COUNTY). From its confluence with Honcut Creek at location T17N R3E Sec. 27, upstream to Highway # 70 Bridge is *Class - B.* Open to dredging from July 1 thrue August 31. From Highway # 70 Bridge to Oroville Dam is *Class - A, closed waters, no dredging permitted at any time.* **FEATHER RIVER, SOUTH FORK (BUTTE and PLUMAS COUNTIES).** The main stem South Fork Feather River from the Oroville Reservoir upstream to Little Grass Valley Dam in Plumas County at location T22N R9E Sec. 31 is *Class - C.* **FEATHER RIVER, MIDDLE FORK (BUTTE and PLUMAS COUNTIES)** is closed to dredging all year from Lake Oroville to the mouth of an unnamed drainage on the east side of the river 3/4-miles downstream of the Milsap Bar Bridge to the mouth of the Nelson Creek. In the remainder of the Middle Fork Feather River, no dredge with the intake larger than four-inches may be used. Middle Fork Feather River from Lake Oroville to the town of Beckworth in Plumas County, mining is prohibited in the *Wild Section of the river.* Check with the USFS for Plumas National Forest Map and for additional regulations covering mining and establishing claims.

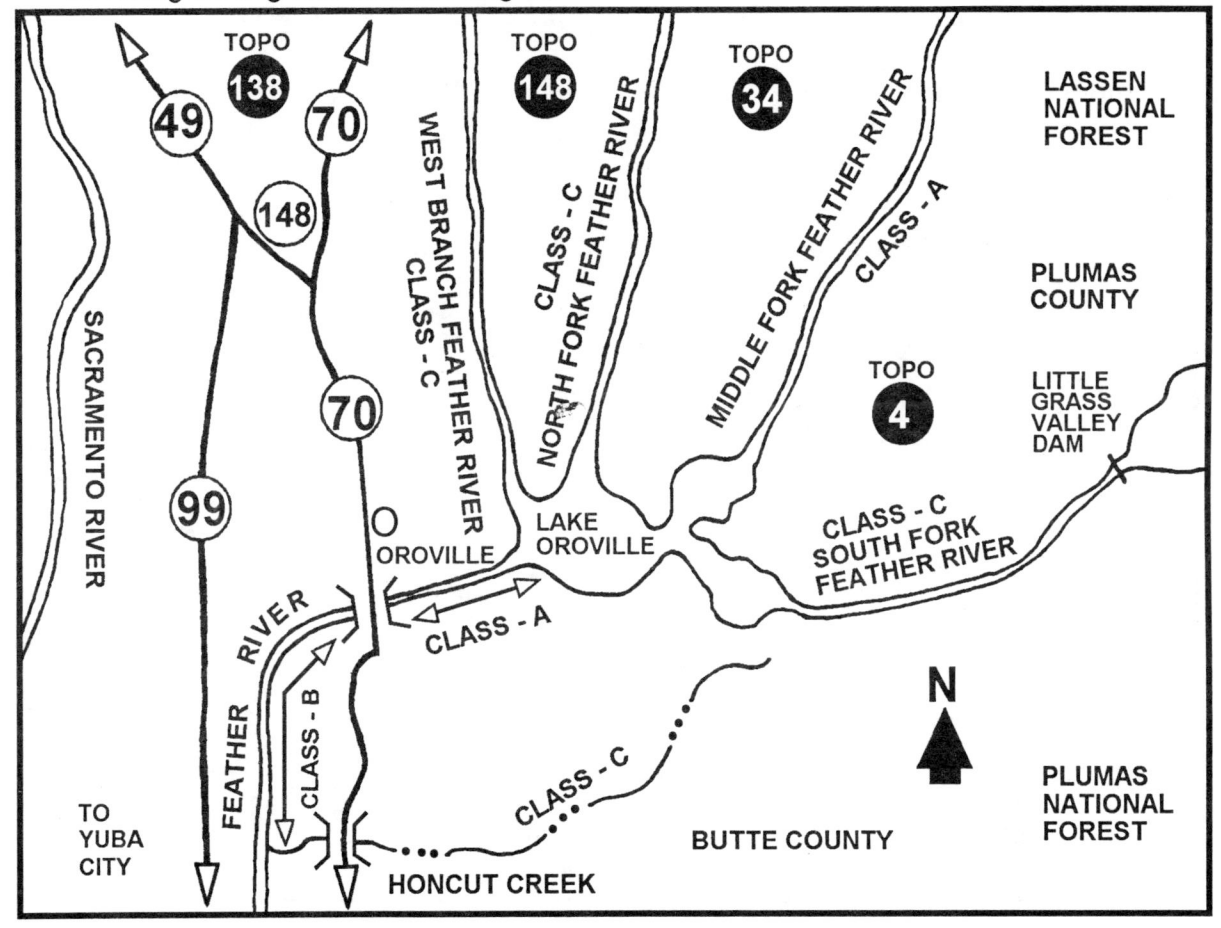

CALAVERAS COUNTY. East of Highway # 49 is *Class - C.* Open to dredging from the fourth Saturday in May thru October 15. The remainder of Calaveras County is *Class - H.* Open all year around except where noted. Use the Stanislaus National Forestry Map, California Highway Map, County and Bureau of Land Management Maps.

A spider-webbed system of Tertiary channels finger their way across Calaveras County from Mokelumne River that borders the north County line, to the south 'n southwest direction to the Stanislaus River bordering the southern County line. Most of the creeks, tributaries and feeder creeks in between its County borders cut thru and across the aged channels that **host the placer gold particles.**

Past hydraulicing cleaned out the concentrations of placer patches and by-passed smaller unprofitable to mine the gravel basins, leaving these to be re-discovered and **placer** mined by today's vacuum-dredgers. Present day erosion continues to move the gravel and relocate the placer gold occurrences into newer riffles where concentrated piles begin. Deep potholes were corroded in the pre-volcanic mantle and proved to be very effective riffles with Tertiary washed gravels filling the basins with rich gold particles. Overlooked intermittent patches of ancient gravel channels holding placer gold can be located with a metal detector along the O'Neil and San Antonio Creeks, (T4N R13E), Topos; *Murphys, Jesus Maria and Esperanza Creeks.* Topo; *Rail Road Flat.*

Bottom gravels adjacent to the Neocene Channels have proved to be very rich in gold particles by past history of mining. Today's modern gravel runs usually are found to be with an intermittent vertical rise 'n fall in depth, with its placer gold being eroded into the adjacent feeder creeks. Check the old tailing piles of any area that has a past rich history in **placer and hydraulic mining.**

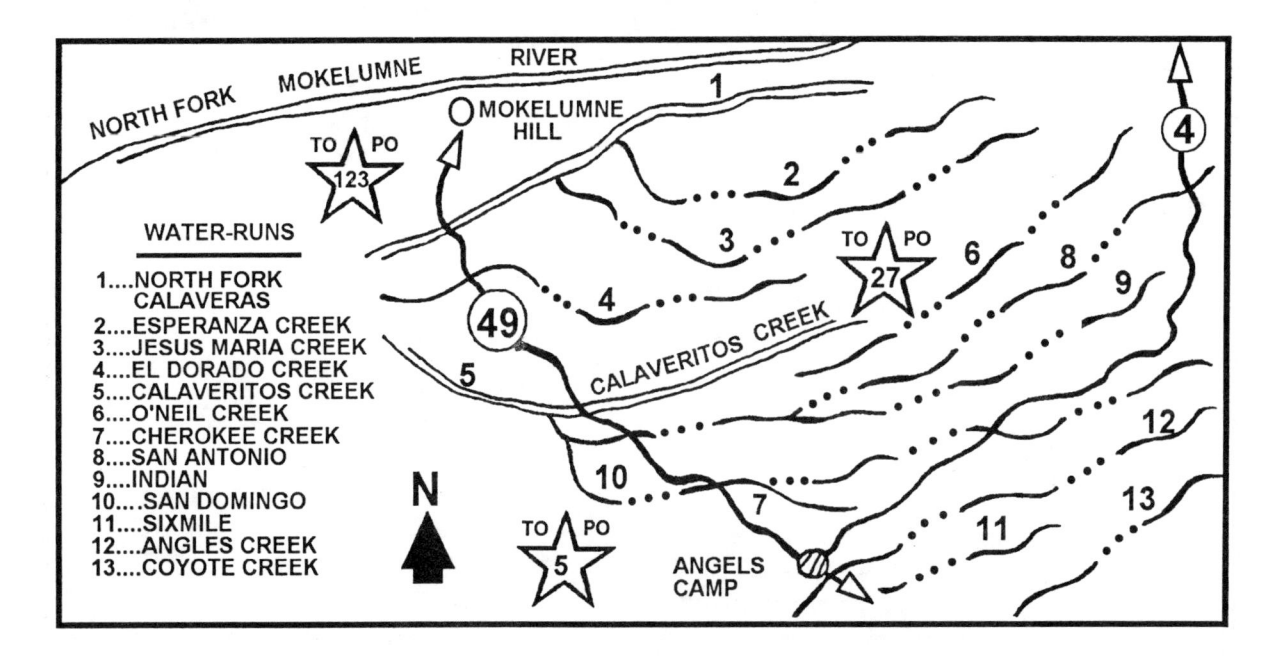

WATER-RUNS

1....NORTH FORK CALAVERAS
2....ESPERANZA CREEK
3....JESUS MARIA CREEK
4....EL DORADO CREEK
5....CALAVERITOS CREEK
6....O'NEIL CREEK
7....CHEROKEE CREEK
8....SAN ANTONIO
9....INDIAN
10....SAN DOMINGO
11....SIXMILE
12....ANGLES CREEK
13....COYOTE CREEK

CALAVERAS RIVER and TRIBUTARIES. (CALAVERAS and SAN JOAQUIN COUNTIES). Below New Hogan Reservoir is *Class - B.* Open to dredging from July 1 thru August 31. East of Highway # 49 in Calaveras County is *Class - C.* Open to dredging from the fourth Saturday in May thru October 15, the remainder of the County is open all year around. San Joaquin County is *Class - H* and is open the year around.

MOKELUMNE RIVER. (CALAVERAS, SAN JOAQUIN and AMADOR COUNTIES). From Lockeford upstream to Pardee Dam is *Class - H.* From Pardee Dam upstream to the Electra Powerhouse is *Class - C.* Open to dredging from the fourth Saturday in May thru October 15. From the Electra Powerhouse to the mouth of the Slaughterhouse Canyon (common corner to Sections 11-12-13-14- T6N R12E is *Class - H,)* open all year. **STANISLAUS RIVER (CALAVERAS, SAN JOAQUIN, TUOLUMNE & STANISLAUS COUNTIES).** From the Santa Fe Railway Bridge upstream to Goodwin Dam is *Class - A.* East of Highway # 49 in Tuolumne County is *Class - C.* Open to dredging from the fourth Saturday in May thru October 15, the re- mainder of Tuolumne County is *Class - H,* open all year around except where noted. Stanislaus River from Melones Dam upstream to Camp Nine Road Crossing is *Class - C.* Stanislaus County is *Class - H,* except where noted.

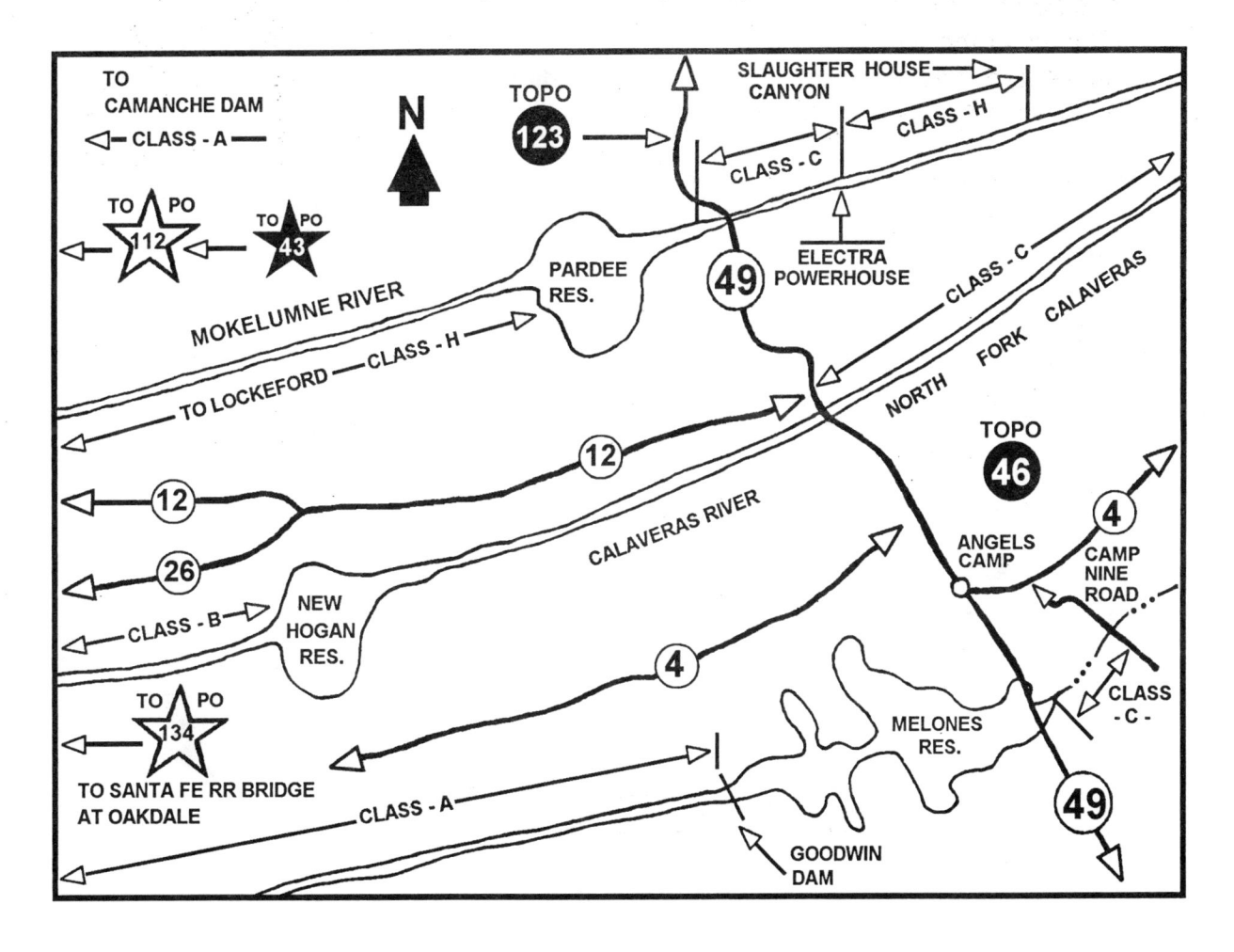

DIAMONDS IN CALIFORNIA PLACER GRAVEL.

Many diamonds in varying grades and sizes have been discovered in several Counties, usually from sluicing, hydraulic and placer mine operations. Thrue the years hydraulic mining before restrictions were placed on those operations, mining methods unknowingly flushed thousands of diamonds into the Sacramento Valley gravel concentrates. Only an occasional diamond was found in the ground sluices manned by the old time miners.

Prospectors, dredgers and gold searchers keep a sharp eye out for the light weighted specific gravity 3.5 diamonds that is easily overlooked and assumed to be quartz, zircon, henacite or some other common crystal by the old time miner.

Diamonds have been found in Counties of BUTTE, AMADOR, DEL NORTE, EL DORADO, TUOLUMNE, PLUMAS, HUMBOLDT, TULARE, TRINITY, SAN DIEGO, NEVADA, CALAVERAS and PLACER COUNTIES.

Most notable discoveries have been found along the Mother Lode Belt where striking reverse faults and shear zones host overlooked and unrecognizable diatremes. Wherever the land is familiar with serpentized shists, serpentinites and peridotoes, or favorable ultrmafic rock protrusions diamonds will be located. Gravels of the water-runs and its stream sediment is where most diamonds will be located, unlike the gravity drop of gold, diamonds will travel with the flow of the stream and settle where the force of the stream dictates.

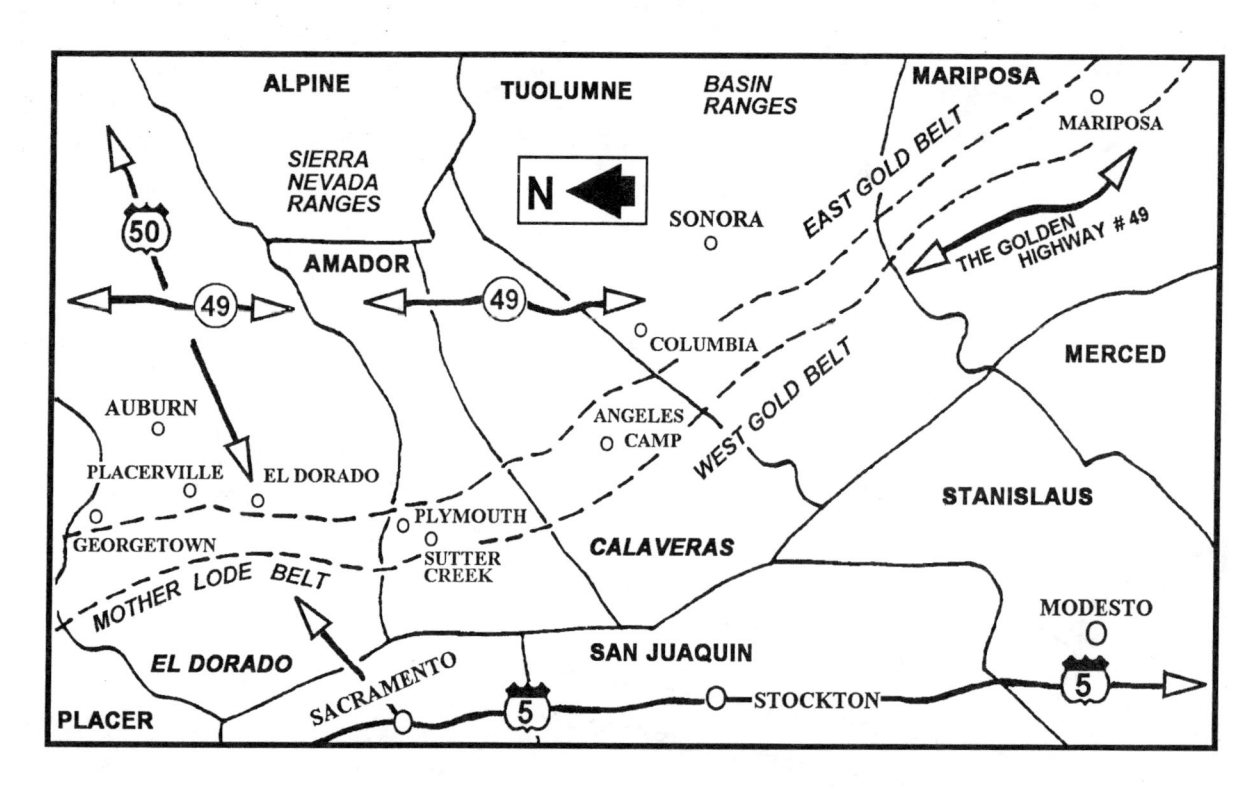

PLATINUM; Where to locate and how to identify its source.

Platinum group metals (PGMs), a six family group are found in territories from Del Norte County down as far as Tulare County on a linear belt of serpentine and peridotite ledges, outcropping from a megashear of ultramafic rock and chromite surface rock clusters of gossans.

When exploring for the PGM group is conducted, the searcher must confine themselves to areas known for PGM occurrences in magmatic results in basic and ultrabasic rocks, and where serpentine, peridotite and greenstone formation exists.

Gravel basins, benches, terraces and high-level ancient channels are best bet for exploration of the PGMs. Some knolls, hills and buttes are capped with old stream beds holding in their Tertiary gravel the gold and Platinum located within the back-country of the Forest Service.

Look to feeder-streams with gravel aggregate amongst the out-back in the Klamath Mountain Range. Platinum has been placered in Beegum Creek located in Tehama County, northwest of Red Bluff alongside of California Highway # 36. Very near to this same area, also with a history of Platinum in its gravel, is the Cottonwood Creek near Platina.

Rivers famous for past gold placers associated with PGMs are the American, Consumnes, Klamath, Trinity, Yuba and all of their tributaries and feeder streams. The dredging fields northwest of Junction City along the Trinity River has a history of being the largest producer of Platinum in California. The entire length of the Trinity River has produced from its gravel placer deposits enormous amounts of gold and platinum particles. The Putah Creek located in Yolo County is a good source for locating Platinum and gold particle occurrences.

Any chromite and nickel lode mine production figure will be a clue to the PGM content of the placer gravel in the adjacent creeks and water-runs that set below the lode mines. The ore dump aprons have been leached by seasonal rains sending the PGM particles percolating their way into the lower tributaries and all of their feeder streams. For further information on PGMs, visit the research department at any large University Library.

PLATINUM MAP LOCATIONS; FROM CENTRAL TO NORTHERN CALIFORNIA WATER-RUNS. *PLATINUM GROUP METALS (PGM)* can be located at bedrock clinging to the crevices and fissures in small particle sizes. In some localities, PGMS did exceed gold in quantity and was discarded by the early '49ers as nuisance concentrates. The serious dredge miners working northern California have Platinum nuggets to show for their work, either in natural nugget form, or as a melt-down button hanging from their neck chains in proud display.

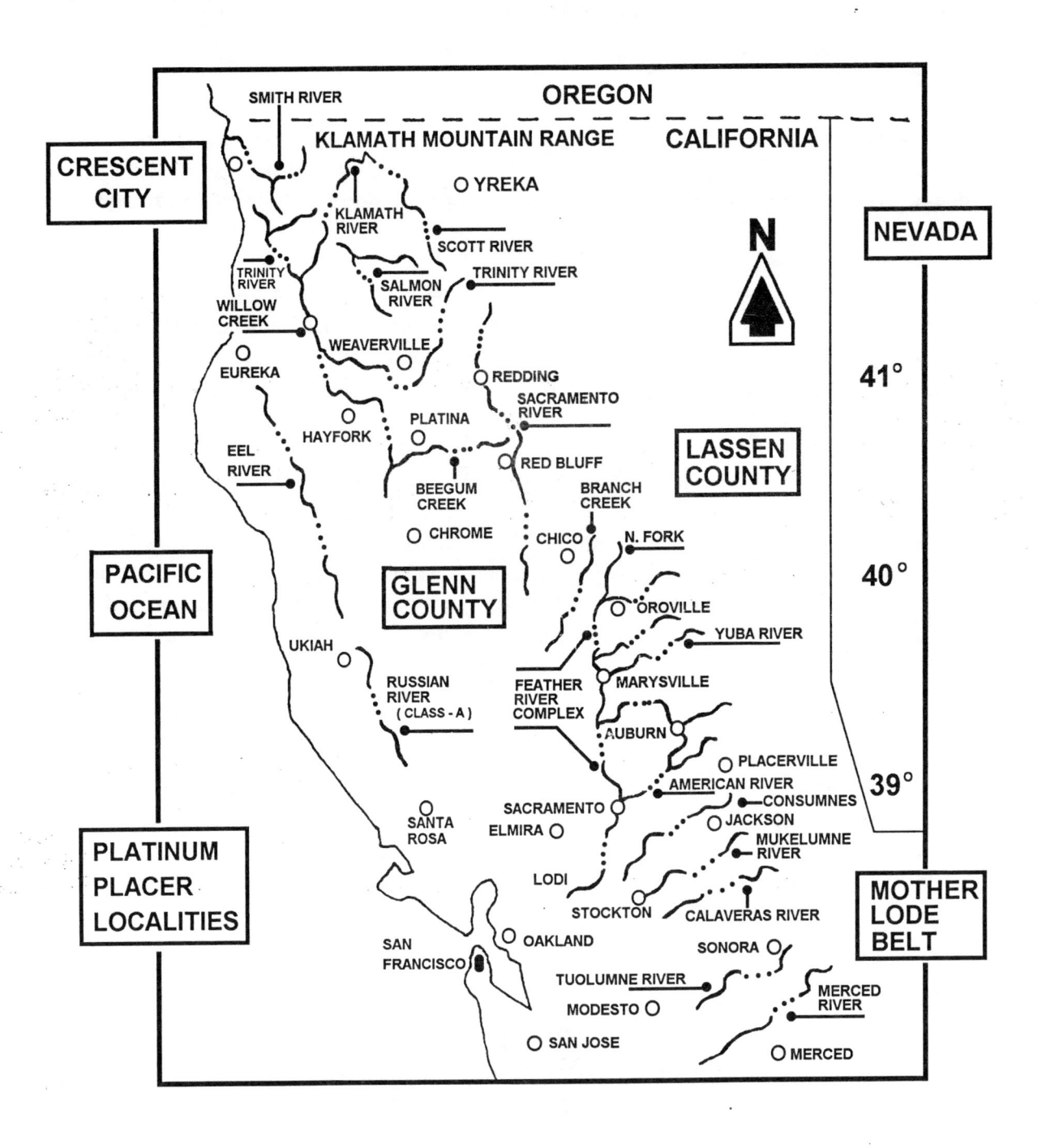

MODOC COUNTY...WARNER MOUNTAINS. Gold sniping with a metal detector country. Modoc National Forest, 441 North Main St., Alturas, CA., 96101. 916-233-5811. Warner Mountain Ranger District, P.O. Box 220, Cedarville, CA., 96103. 916-279-6116. Modoc County is located in the northeast corner of California with its history of gold production centered between the towns of Adin and Canby along State Highway # 299, and limited mining operations in the northern part of the Warner Mountains just off of either side of Highway # 395.

Placer gold interest took a back seat to lode mining operations mainly because gold showed up on the surface from outcroppings, more than the watercourse gravels proved out. Old mine dumps are prime suspect for rare earth metals, noble and exotic minerals with elements that are overlooked fortunes. Chemical Analysis Testing will allow the searcher to determine what the old mine dump contained and what its worth would be. A simple five minute step-method testing sequence is all that would be required.

Possibilities are favorable at higher elevations in the Warner Mountains where gravels in the Bidwell Creek, Pine Creek and their tributaries and with feeder creeks draining the area the year around, may have virgin (overlooked) placer gravel deposits. The higher elevations of the Warner Mountains have seen very limited prospecting activities, and it is in this area that holds promise of gold particles, rare-earth metals, elements of exotic minerals. Gold-bearing outcroppings, ledges and scattered gossans with oxidized pyrites could crest the many ridges with indications where overlooked fortunes may be lying beneath the talus and pine covered countryside. Down-hill from the old mining activities are ideal locations for sniping for gold nuggets with a metal detector. Be alert to existing mining claims and private property. Check with the County Recorders Office for land status in the area of interest.

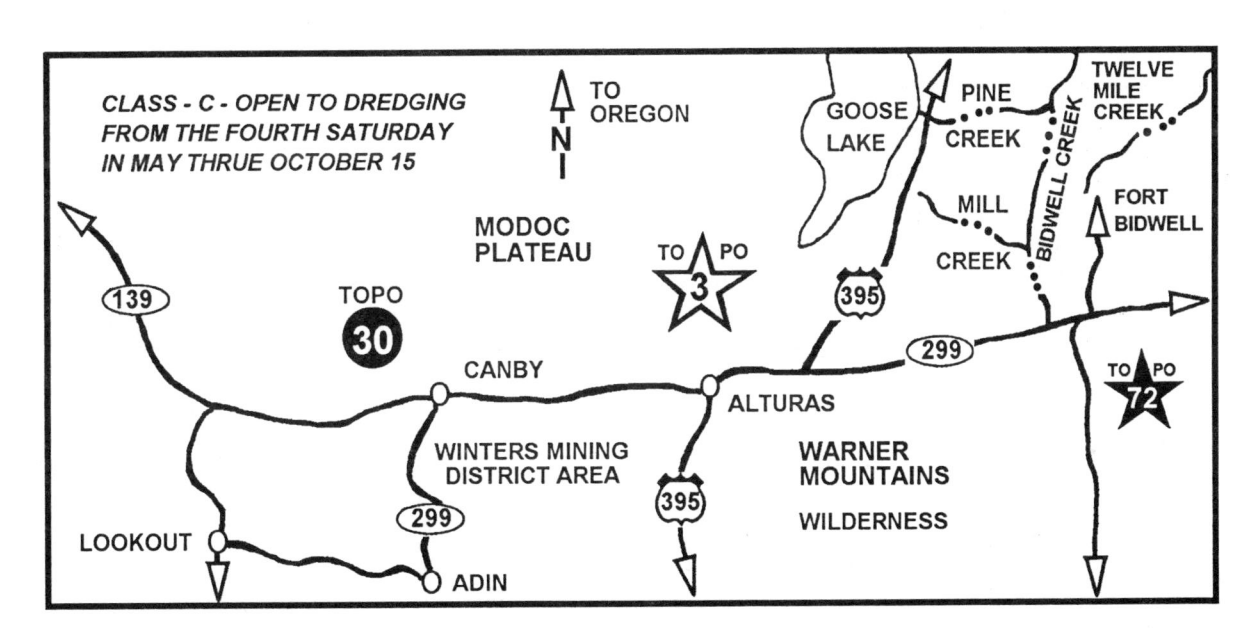

EEL RIVER, MAIN STEM, (HUMBOLDT, TRINITY AND MENDOCINO COUNTIES). Humboldt County is *Class - E.* Open to dredging from July 1 thrue September 30. Trinity County is *Class - E.* Open to dredging from July 1 thrue September 30. Mendocino County is *Class - A.* Closed waters to dredging, and tributaries are included. **EEL RIVER, MIDDLE FORK AND TRIBUTARIES. TRINITY AND MENDOCINO COUNTIES.** From its mouth at Dos Rios upstream, including all tributaries for one-mile, of their full length from their confluence with the Middle Fork EEL River is *Class - A,* closed.

TRINITY RIVER. (HUMBOLDT AND TRINITY COUNTIES). Main stem below Lewiston Dam. The main stem Trinity River from the Klamath River upstream to the South Fork Trinity River is *Class - A, (closed.),* From the South Fork Trinity River upstream to the North Fork Trinity River is *Class - H, (open all year around).* From the North Fork Trinity River upstream to Grass Valley is *Class D, (open to dredging from July 1 thrue September 15).* From Grass Valley upstream to Lewiston Dam is *Class - A, (closed).* **TRINITY RIVER, NORTH FORK AND TRIBUTARIES (TRINITY COUNTY).** The North Fork Trinity River and its tributaries upstream from Hobo Gulch Campground are *Class - A, (closed).*

DEER CREEK, (NEVADA COUNTY). From Ponderosa Way below Rough and Ready Falls at location T16N R7E Sec. 13, upstream to Highway # 49 and east of Highway # 49 is *Class -C, (open to dredging from the fourth Saturday in May thrue October 15, except where noted).* **VAN DUZEN RIVER, HUMBOLDT COUNTY.** No dredging with intake larger than six inches may be used. Humboldt County is *Class - E, (open to dredging from July 1 thrue September 30).* Use Trinity, Humboldt, Mendocino County Maps, California Highway Map, Six River National Forest and Mendocino National Forestry Maps for location.

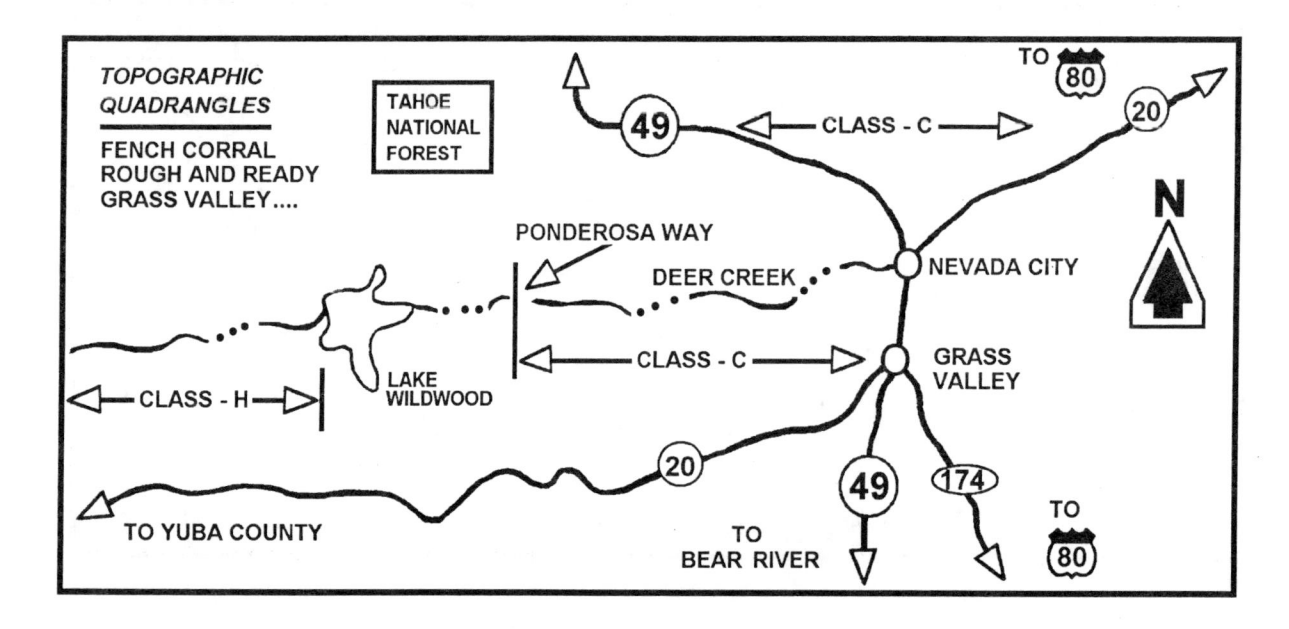

TOPOGRAPHIC QUADRANGLES
FENCH CORRAL
ROUGH AND READY
GRASS VALLEY....

CHURN CREEK AND TRIBUTARIES, (SHASTA COUNTY). From the junction with Newton Creek and the meeting of Highway I-5 at location T32N R5W Sec. 7, upstream is *Class - C,* open to dredging from the fourth Saturday in May thru October 15. **CLEAR CREEK, (SHASTA COUNTY).** From the junction with the Sacramento River at location T31N R5W, upstream to the McCormick-Saeltzer Dam is *Class - B.* open to dredging from July 1 thrue August 31.

FLAT CREEK, (SHASTA COUNTY). With location at T32N R5W Sec. 9. Open to dredging throughout the year as *Class - H.*
MOODY CREEK AND TRIBUTARIES, (SHASTA COUNTY). Tributary to Stillwell Creek.

ROCK CREEK, (SHASTA COUNTY). From its junction with the Sacramento River at location T32N R5W Sec. 21, upstream is *Class - H,* open to dredging all year long. **SACRAMENTO RIVER, (SHASTA AND TEHAMA COUNTIES).** From the Squaw Hill Bridge (between Corning and Vina), upstream to Keswick Dam at location T32N R5W Sec. 21 is *Class - A.* Closed waters.

SALT CREEK, (SHASTA COUNTY). At location T32N R5W below Keswick Dam, downstream from Highway # 299 bridge is *Class - A.* Closed waters.
SULFUR CREEK AND ITS TRIBUTARIES (SHASTA COUNTY). From its junction with the Sacramento River at location T32N R5W Sec. 25, to the SP RR tracks is *Class - A.* Closed waters.

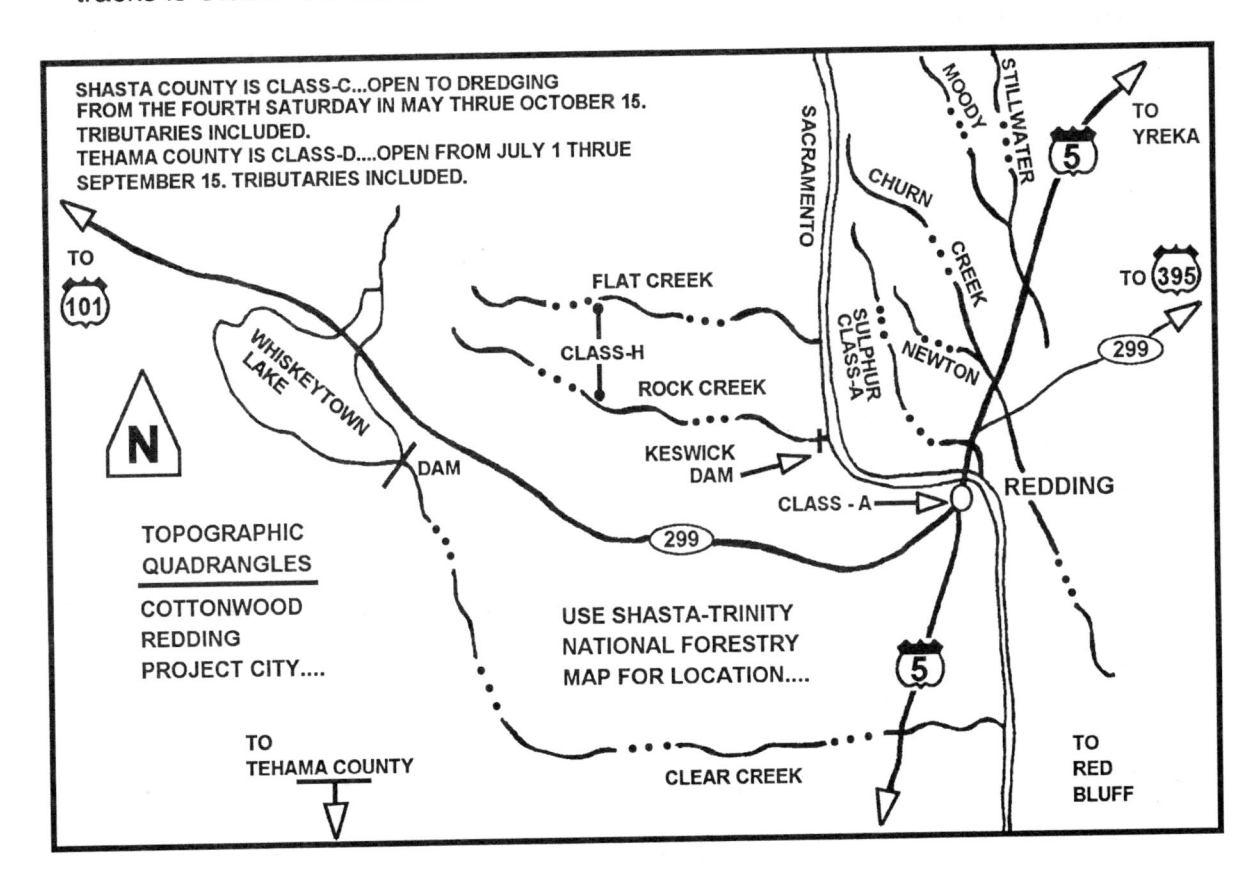

110

PLUMAS COUNTY IS *Class - C,* except where noted. Open to dredging from the fourth Saturday in May thru October 15. Use the Lassen National Forest Map, Plumas National Forest Map, County and California Highway Map for location.

Huge amounts of **placer gold** had been mined during the historical **gold** rush from the Feather River and its Territory and Recent age feeder creeks. Today's dredgers are intermittently active during seasonal placer mining. Present undiscovered surface ore outcrops and scattered high-grade pockets contribute to placer gold deposits in the adjacent feeder creeks of the **back-country** during the seasonal run-off.

The La Porte channel, (considered to be the main Tertiary Channel of the ancient Yuba River), continues to be disturbed and uplifted by intermittent faulting, creating re-arrangement of the hidden quartz-rich gravel deposits. Winter and summer thunder storms rake the newly exposed Terrace gravels, holding basins and ancient channels sending the gathered **gold particles** into the adjacent tributaries and feeder creeks.

Some of the richest, mineralized surface and deeply seated gravel deposits are contained within the Plumas-Eureka State Park borders (T22N R11E). The water-runs adjacent to the Park are affected by seasonal run-off carrying **placer gold particles** from the inside of the Park. The Meadow Valley district plays host to Recent and Pleistocene Valley alluvium containing placer gravel gold particles, (T24N R8E). Check the hydraulic piles 'n heaps for overlooked gold nuggets 'n particles. Be alert to mining claims and private property in the gold belt of the Northern Sierra Nevada area.

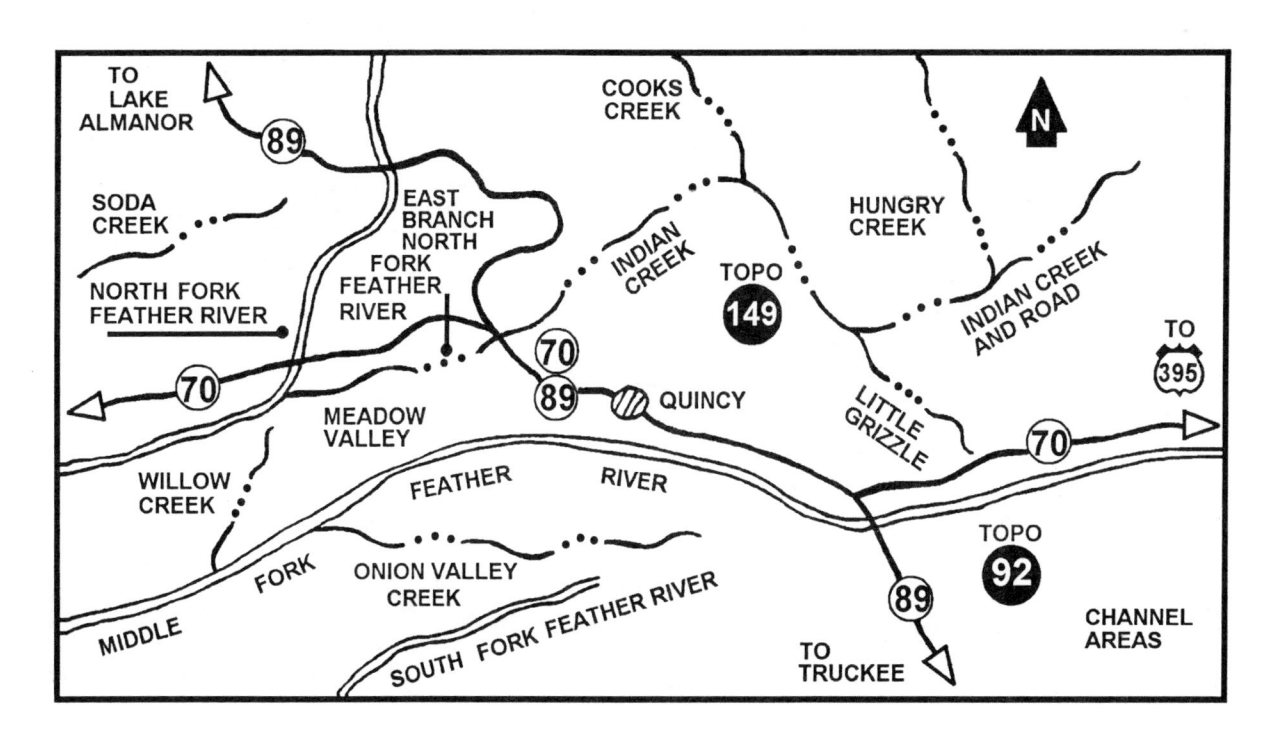

SUTTER CREEK, AMADOR COUNTY. The main stem Sutter Creek from Highway #49 upstream to Pine Gulch Road is *Class - H.* Amador County east of Highway # 49 is *Class - C,* the remainder is *Class - H.* Topo; *Amador City,* T7N R11E Sec. 31. **TUOLUMNE RIVER, (STANISLAUS COUNTY).** The main stem Tuolumne River from the Waterford Bridge upstream to La Grange Dam is *Class - A,* the remainder is *Class - B,* open to dredging from July 1 thru August 31 except where noted. The Fish and Game Department consistently change these dates and classification...**CHECK! TURNBACK CREEK AND ITS TRIBUTARIES are *Class-A.* (TUOLUMNE COUNTY)** east of Highway # 49 is *Class - C,* the remainder of the county is *Class - H.* Use Topo; *Standard and Tuolumne,* and Stanislaus National Forestry Map for location. **YELLOW CREEK INCLUDING TRIBUTARIES, (PLUMAS COUNTY).** Dredge with an intake larger than four-inch is prohibited. *Class - C,* open to dredging from the fourth Saturday in May thru October 15, except where noted. Junction with the North Fork Feather River at location T25N R6E Sec. 24 , Plumas National Forestry Map. Use Topo Quadrangles, *Caribou and Humbug Valley* for location. **NELSON CREEK INCLUDING TRIBUTARIES, (PLUMAS COUNTY).** The main stem Nelson Creek is *Class - C,* open to dredging the fourth Saturday in May thru October 15. Junction with the Middle Fork Feather River at location T23N R10E Sec. 16 in the PlumasNational Forestry Map and on the Topo Quadrangle Map *Blue Nose MTN.* **SMITH RIVER MIDDLE FORK , DEL NORTE COUNTY.** The main stem Smith River Middle Fork is *Class - D,* open to dredging from July 1 thru September 15, except where noted. Del Norte County is *Class - E,* open to dredging from July 1 thru September 30 except where noted. Dredge with an intake larger than six-inches is prohibited.

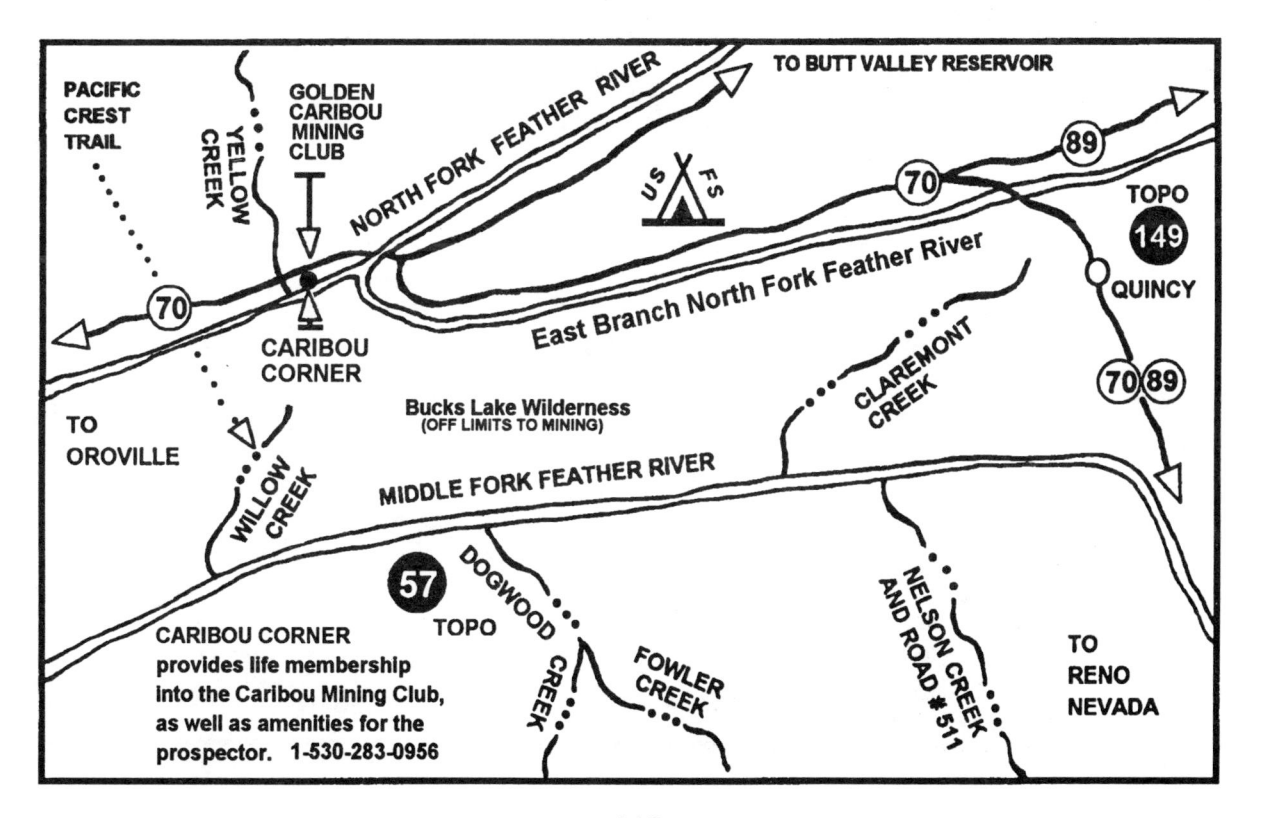

112

MARIPOSA COUNTY. Within the external boundaries of the National Forest is *Class - C,* open to dredging from the fourth Saturday in May thru October 15, the remainder of the County is *Class- H,* open all year around except where noted. Use the Stanislaus National Forestry Maps, County and California Highway Maps along with the Bureau Of Land Management (BLM) Maps for location. Most water-runs of Mariposa County carry placer gold particles which come from the Mother Lode Gold Belt Area influence made up from the ancient Terraces, gravel basins and old channels of the Mariposa Formation (Upper Jurassic), serpentine and Greenstone with massive quartz veins in the formations. The ore contains **free gold**, pyrite and arsenopyrite, which is often associated with high-grade ore. The Merced River with its many tributaries, feeder creeks and its scraping freshets crisscross the County, washing and percolating the ancient channel gravels and depositing the placer gold into the lower riffle basins. Good placer take has been reported along the North Fork Merced River and its adjacent feeder creeks. Particle locations are found along the Middle Fork Chowchilla (T6S R20E), Bull Creek (T2S R19E), Merced River (T3S R19E), Mariposa and tributaries (T6S R18E), Owens Creek and its feeder creeks (T6S R17E), Sherlock, Lyons Gulch and the Bear Valley influence (T4S R17E). Wherever the old timers placered, stoped, drift and carried on hard rock lode mining activities, the present day gold searcher stands a better than even chance of locating undiscovered gold-bearing gravels in ancient channels where modern day creeks junction. Check the area placer locations at Greeley (T2S R17E), Kinsley (T2S R18E), Hunter Valley T3S R16E, Bear Valley (T4S R17E), El Portal (T3S R20E), Mount Bullion T5S R17E, Jerseydale (T5S R19E), Bootjack (T5S R19E), Catheys Valley T6S R17E, Whitlock and Whiskey Flat, Sherlock Creek and Colorado (T4S R18E).

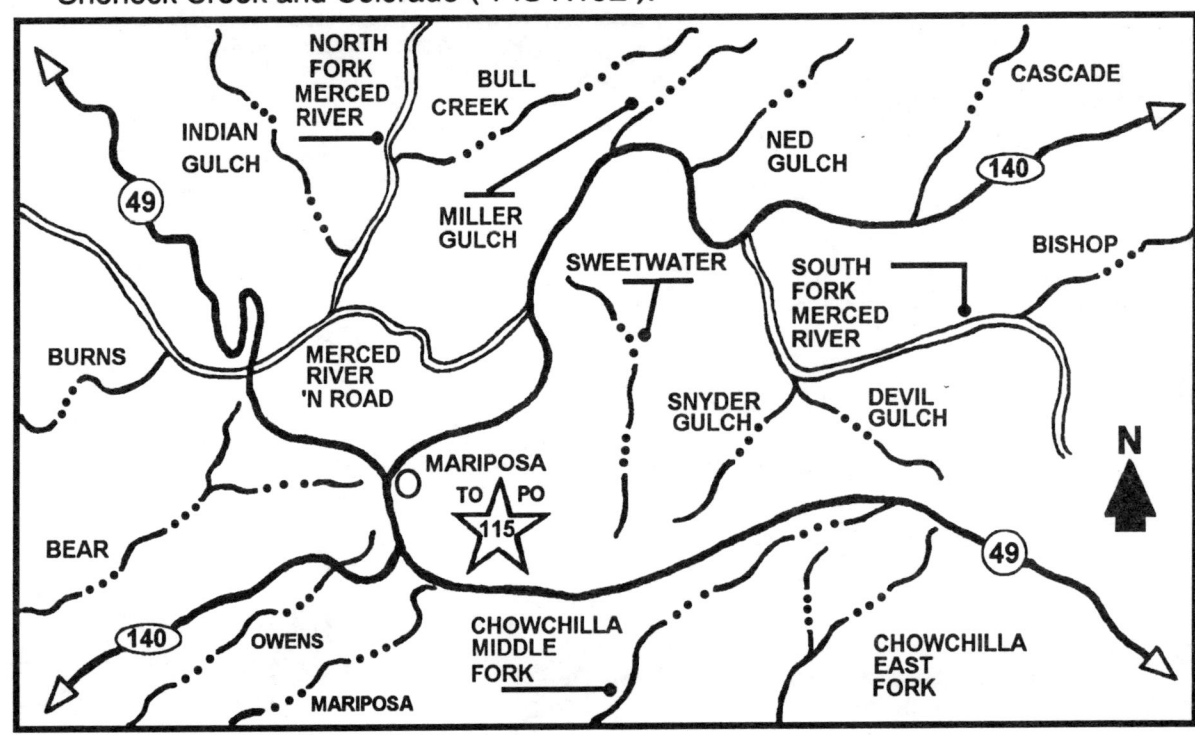

BUTTE COUNTY, has an extensive area of placer deposits that were mined at the old towns of Bangor and Wyandotte in the 1850's. Tertiary gravels cover a broad area and it is believed to be a replica of the ancient delta of the Tertiary Yuba River with evidence of shore gravels of the Pleistocene age. The channel gravels are well rounded, fully cemented and consist of intrusive and metamorphic rock fragments of quartz. Gold is of flaky and rusty to coarse nugget combination.

PLUMAS COUNTY. In recent years vast amounts of placer gold has been recovered from the Feather River complex. Weekend prospectors continue to be active in the many streams of the back-country. The area is underlain with a series of metamorphosed sediments of ancient age. Gold-quartz occur in brecciated zones in the Greenstone and Serpentine conglomerates.

Placer gold can be located in just about every water-run of the area within the back-country. Practically every pan will show some color of sorts. On the Feather River just above the Thermalito Bridge, "color" is consistently located by the week-enders and vacation people. The Milsap Bar campground, on the Middle Fork of the Feather River, can be located about a 30-mile drive from Brush Creek at location T21N R6E Sec. 2, (strike out from Brush Creek work center on USFS road no. 22N62).

The French Creek area just above the Brush Creek work center at approximate location T22N R5E Sec.24 is noted for its placer gold content. Take the Oro-Quincy Highway, USFS road no. 27561 (Moortown Four Mile Ridge Road) to the Mountain House Creek area and then drive the Four-Mile Ridge Road and follow the signs to the area for *panning for the gold.*

A four-mile stretch situated at the confluence of the North Fork and the East Branch of the Feather River at the mouth of the Yellow Creek, between Belden Town and Rich Bar is noted for its placer gold content. Belden Town location at T25N R6E Sec. 24. Rich Bar location at T25N R7E Sec. 21 NE1/4. Use the Plumas National Forestry Map for location.

Contact the Park Headquarters for the Lake Oroville State Recreation Area Campsite reservations at 916-538-2200. *How to pan for gold, where to go for the gold and for gold panning demonstrations and instruction,* contact the Lake Oroville Visitor's Center for reservations and time schedules 916-538-2219. Dredging permits may be obtained at the Oroville Forestry Office.

Stay within the confines of the Milsap Bar Campground and other campsite corridors when engaged in recreational gold panning while on the Middle Fork Feather River. This river has been designated as a National Wild and Scenic River.....where gold prospecting is prohibited.

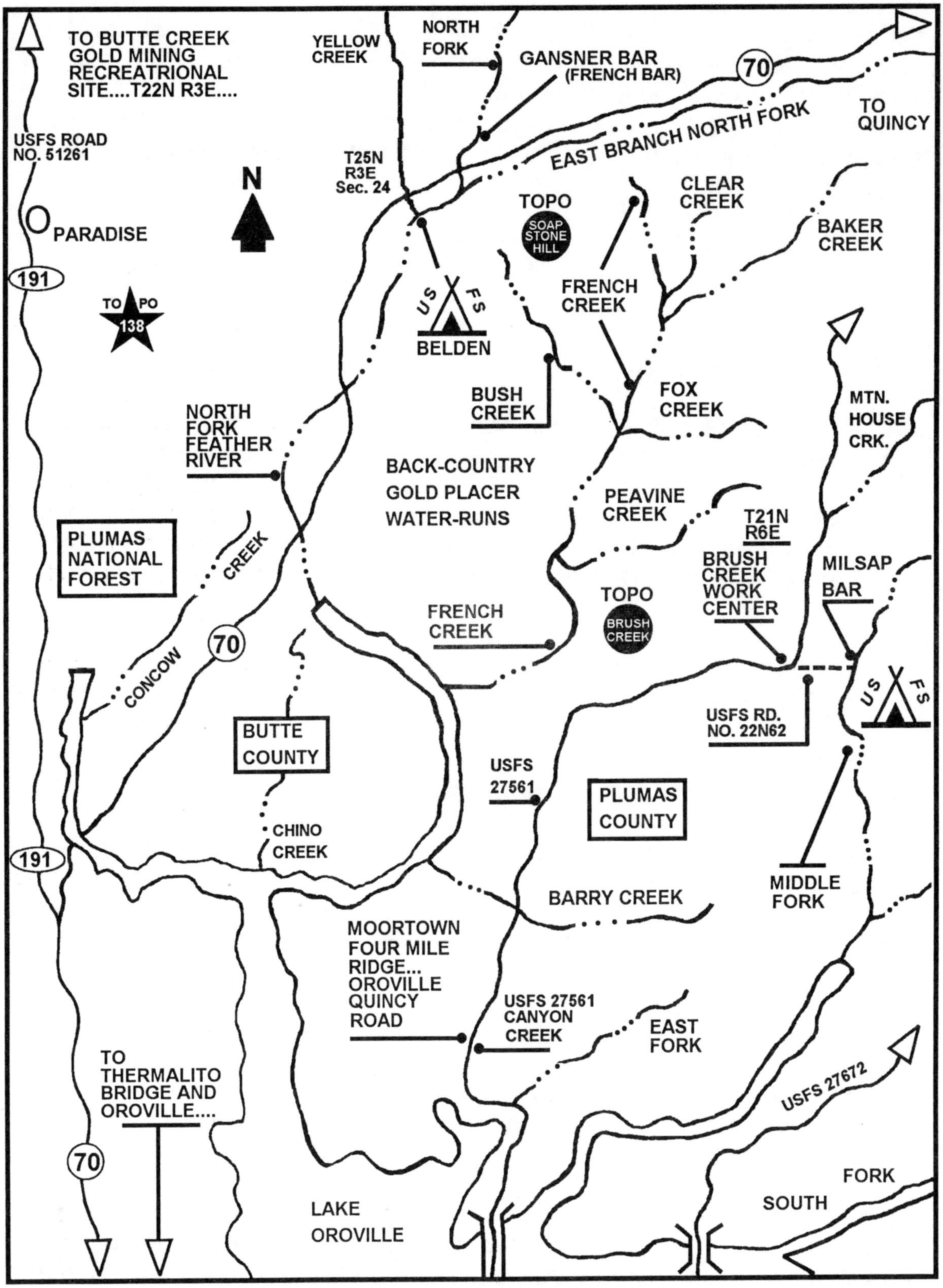
TO BUTTE CREEK GOLD MINING RECREATRIONAL SITE....T22N R3E....
YELLOW CREEK
NORTH FORK
GANSNER BAR (FRENCH BAR)
70
TO QUINCY
EAST BRANCH NORTH FORK
USFS ROAD NO. 51261
N
T25N R3E Sec. 24
PARADISE
191
TOPO
SOAP STONE HILL
CLEAR CREEK
BAKER CREEK
TO PO 138
US FS
BELDEN
FRENCH CREEK
BUSH CREEK
FOX CREEK
MTN. HOUSE CRK.
NORTH FORK FEATHER RIVER
BACK-COUNTRY GOLD PLACER WATER-RUNS
PEAVINE CREEK
T21N R6E
PLUMAS NATIONAL FOREST
CREEK
BRUSH CREEK WORK CENTER
MILSAP BAR
TOPO
BRUSH CREEK
FRENCH CREEK
70
CONCOW
USFS RD. NO. 22N62
BUTTE COUNTY
USFS 27561
PLUMAS COUNTY
CHINO CREEK
191
BARRY CREEK
MIDDLE FORK
MOORTOWN FOUR MILE RIDGE... OROVILLE QUINCY ROAD
USFS 27561 CANYON CREEK
EAST FORK
TO THERMALITO BRIDGE AND OROVILLE....
USFS 27672
70
LAKE OROVILLE
SOUTH FORK

SANTA LUCIA RANGE, (MONTEREY COUNTY). Class - A. No dredging at any time. Los Burros Mineral District is within the Los Padres National Forest as described in the California Division Of Mines, County Report No. 5. The publication can be reviewed at most large University research libraries along with other gold locations for California. *Gold panning, ground sluicing with a rocker, gold sniping with a metal detector and dry washing where applicable are recommended gold mining activities*. Past gold mining history of the area has proved fruitful. Some areas remained inaccessible to prospectors and the old timers left the area virtually untouched and unmined. Several productive bench gravels occur in the South Fork of the Willow Creek, near the junction of the North Fork of the Willow Creek with Dogvine Creek. Other bench gravels with *placer gold deposits* occur on the South Fork of Willow Creek from its headwaters to its junction with Willow Creek. Gravels in skip locations occur in Sections 33, 34, 35, T23S R5E and within Sections 3 and 4 at location T24S R5E. Check those tailing piles. *Placer gold* was located in Dogvine Creek just above the junction with the North Fork of Willow Creek at Section 22, 23, 27 at location T23S R5E. Alder Creek and its feeder creeks flow south past the Mansfield site where the best *gold placer* location of the area exists at location T23S R5E Sec. 35 and 36, T24S R5E Sec. 1 and 2. Plaskett Creek empties into the Pacific Ocean and from this point about one-mile upstream, the placer gold has been located at location T23S R5E Sec. 19 and 20. Check those existing old tailing heaps and piles for overlooked nugget content. Explore the back-country for overlooked hidden stream courses hosting ancient gravel aggregate. Salmon Creek at location T24S R6E Sec. 17/20 has spotty *gold-bearing* gravel with rough access to the west bank where old sluicing has occurred. Check with the Ranger Station located one-mile north of Plaskett Creek for land status.

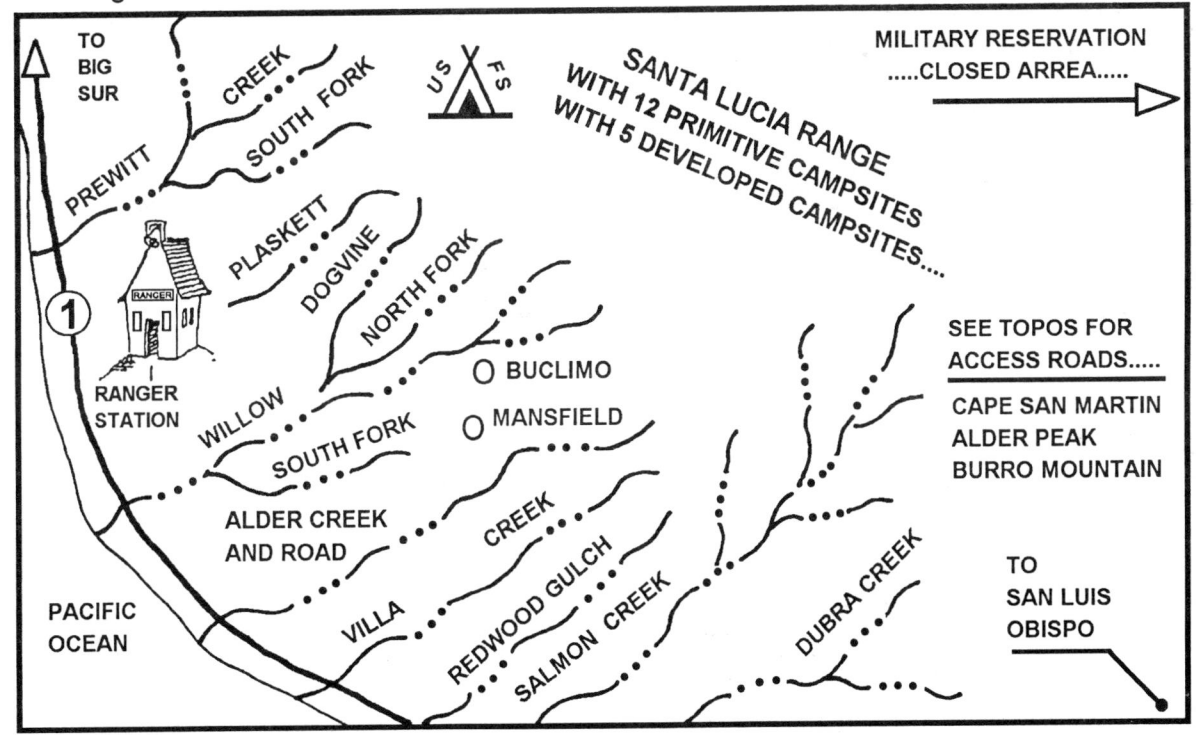

FRESNO AND MADERA COUNTY *are Class - C,* open to dredging from the fourth Saturday in May thru October 15. Within the external boundaries of the National Forest is **Class - C,** the remainder is **Class - H.** Kings River Special Management Area has been closed to vacuum-suction dredging by the USFS. Contact Sequoia USFS for more information.

On the upper San Joaquin, Chowchilla, Fresno Rivers, including their tributaries, small amounts of placer gold had been gathered in the early years. Most of the gold that had been harvested came from the deterioration of Pyritic masses from the adjacent granite over eon-time in accumulation within the river systems. Ancient channel influence of the East Gold Belt of the Mother Lode influence stopped short of the Madera County line. Fresno and Madera Counties are located at the very edge of the mineral influences and contain, in effect, the leaching mineral residues that settled at the edge of the foothills in the area over eons in time.

Small amounts of free gold, quartz gold, stringers in veins are in limited locations and have remote influence on the adjacent gravels of Fresno and Madera County water-runs as downhill seasonal run-off. The flood of 1862 wiped out most of the gold-bearing gravel that had been laid down by adjacent ancient channels where hydraulic and drag line dredging had been working, leaving no trace of the tailing piles in and around the major rivers and their tributaries. Gold that is obtained by dredging the river gravels today, come from small gravel deposits that have remained hidden and undiscovered. In the Frian area gold is fine to flaky and occurs in the small gravel Terrace deposits adjacent to the San Joaquin, Chowchilla and Fresno rivers.

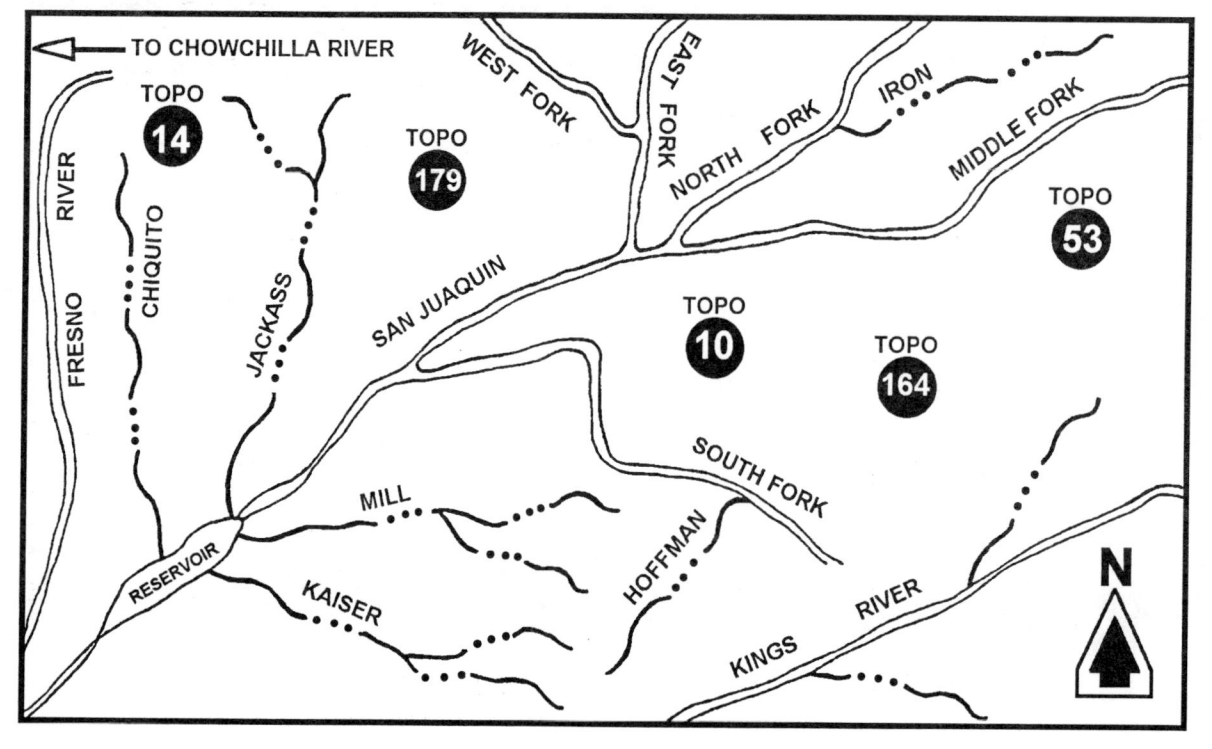

DOGTOWN (BRIDGEPORT--VIRGINIA CREEK). Use Topo. Quadrangle *BIG ALKALI* at location *T4N R25E Sec. 27.* Use MONO County Map, California Highway Map and BLM Surface and Mineral Status Map for Bridgeport-Bodie Hills Mono Lake Areas. (Write for BLM Index Map).

MINERAL WITHDRAWAL GOLD MINING RECREATIONAL CORRIDOR. OPEN TO THE PUBLIC AS A FREE-USE SITE.

Bureau of Land Management Free-Use Site is located at approximately seven-miles south-east of Bridgeport opposite the turn-off on Highway # 395 into the Bodie State Park. Dogtown site is located on the steep west down-side of Highway # 395 and takes in the entire Section 27 for the Free-Use Site.

Tailing piles are visible from the Highway # 395 along with any dredgers who may be working Virginia Creek. Much of the Section 27 of Dogtown is out of sight from Highway # 395 behind the brush, trees and hills, leaving a great deal of Section 27 available for working.

Plenty of old miner tailing piles on Virginia Creek, Dog Creek and Dunderberg Creek. The old miners lost about 50% of their placer workings due to their inefficient mining methods. Check 'n snipe the tailing piles 'n heaps with a good dependable metal detector, (gold detector). Low volume flow of water in the creeks during the midsummer period. Be alert to existing mining claims and private property surrounding the perimeter of the Free-Use Site of Section 27.

For more detailed information write to; BLM California State Office , Federal Office Bldg., 2800 Cottage Way, Sacramento, CA., 95825. 916-978-4754.

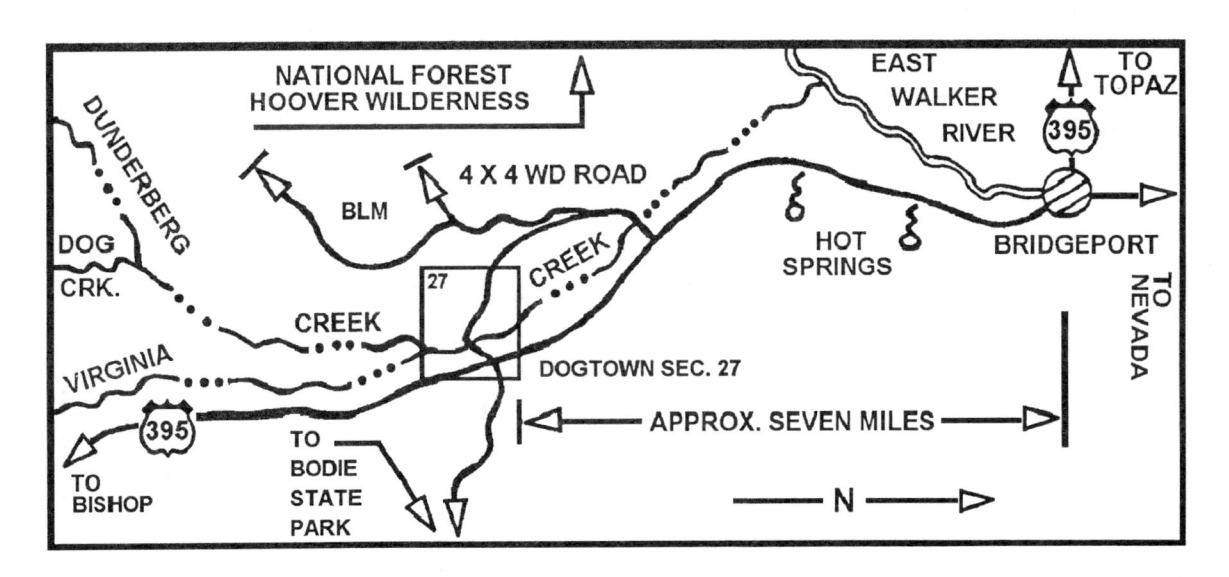

KEYSVILLE -- KERN COUNTY-- LAKE ISABELLE. KERN County is *Class - H* open all year around except where noted. KERN River and its tributaries (KERN and TULARE Counties), from Isabelle Dam upstream is *Class - A, closed waters.*

MINERAL WITHDRAWAL RECREATIONAL GOLD MINING CORRIDOR. OPEN TO THE PUBLIC AS FREE-USE SITE. TWO RECREATIONAL GOLD MINING SITES ARE AT T26S R32E Sec. 25 SW 1/4, Sec. 36 N 1/2, NE 1/4, SE 1/4. Use KERN County road map, California Highway Map, National Forest Map, BLM Surface-Mineral Management Status Map and the Topo Quadrangle *Alta Sierra map.*

Location is forty miles NE of Bakersfield on Highway No. 178 to Keysville (Lake Isabelle). From the off-ramp of California Highway # 178 at Lake Isabelle, Sec. 36, can be reached quite easily. Keysville Road travels thrue the Free-Use Site at Sec. 25. Full services in the area.

Gold panning, rocker-sluicing, dry washing and dredging with a maximum 3" intake is permissible. No sluice-riffle boxes or dry-washer with collecting areas greater than six square feet are permitted. No mechanized earth moving equipment is permitted. Snipe the bedrock and crevices for the gold.

All areas that are disturbed by recreational gold mining must be replaced to its similar (original) condition prior to mining activity. Highbanking is prohibited. When working in the Kern River, all dredging activities must be at least 100 feet apart from the next operator. A valid dredging permit from the California Fish and Game must be available upon request. For further information write to; BLM Resource Office, 4301 Rosedale Highway, Calif. 93308.   805-861-4236.

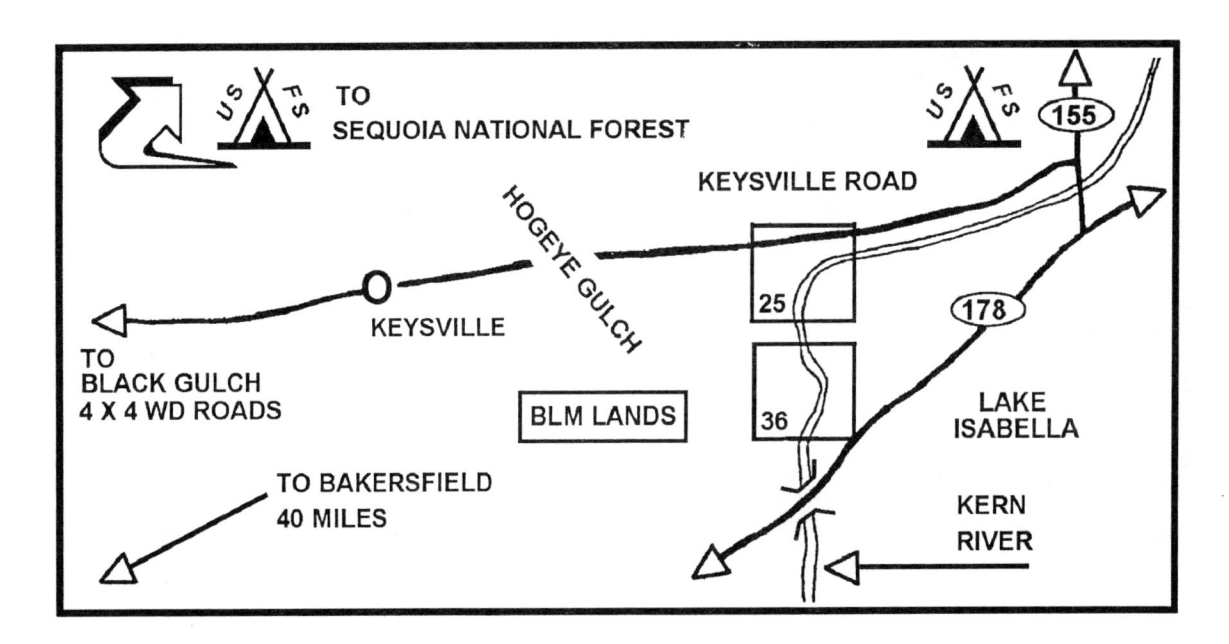

MERCED COUNTY is *Class - H,* open all year around except where noted. Merced River, (Merced County), the main stem Merced River from the San Joaquin River upstream to Crocker-Huffman Dam (upstream from Snelling) is ***Class - A closed waters.*** **MERCED RIVER , (MARIPOSA COUNTY),** the main stem Merced River is ***Class - C,*** open to dredging from the fourth Saturday in May thru October 15. The main stem North Fork Merced River (Mariposa County) is ***Class - C,*** open to dredging from the fourth Saturday in May thru October 15. A flood plain was created by the run-off from the leaching edges of the West Gold Belt where the gold particles were dispersed into concentrations within the Merced River and its feeder creeks. These same conditions were found to exist on the Merced Rive between Snelling and the Merced Falls and was worked by seven commercial drag-line dredges and stackers up until 1952 in an area nine-miles long by 1/2-mile to 1 1/2 miles wide. Most of the gold-bearing gravel of any importance was placer mined at this location, with gold exceedingly flaky and fine with a small amount of **platinum** present.

The feeder creeks adjacent to the Merced River in Merced County deserve attention in as much that they too were influenced by scattered and overlooked **gold-bearing** gravel, that had become impregnated by the leaching qualities from the West Gold Belt presence. The best reported placer 'n dredging field was inundated by Lake McSwain and Lake McClure reservoirs just over the county line in Mariposa on the Merced River. This area was an extension of the same influential **gold-bearing** gravels that were present on the Merced River and its many tributaries, between Snelling and Merced Falls.

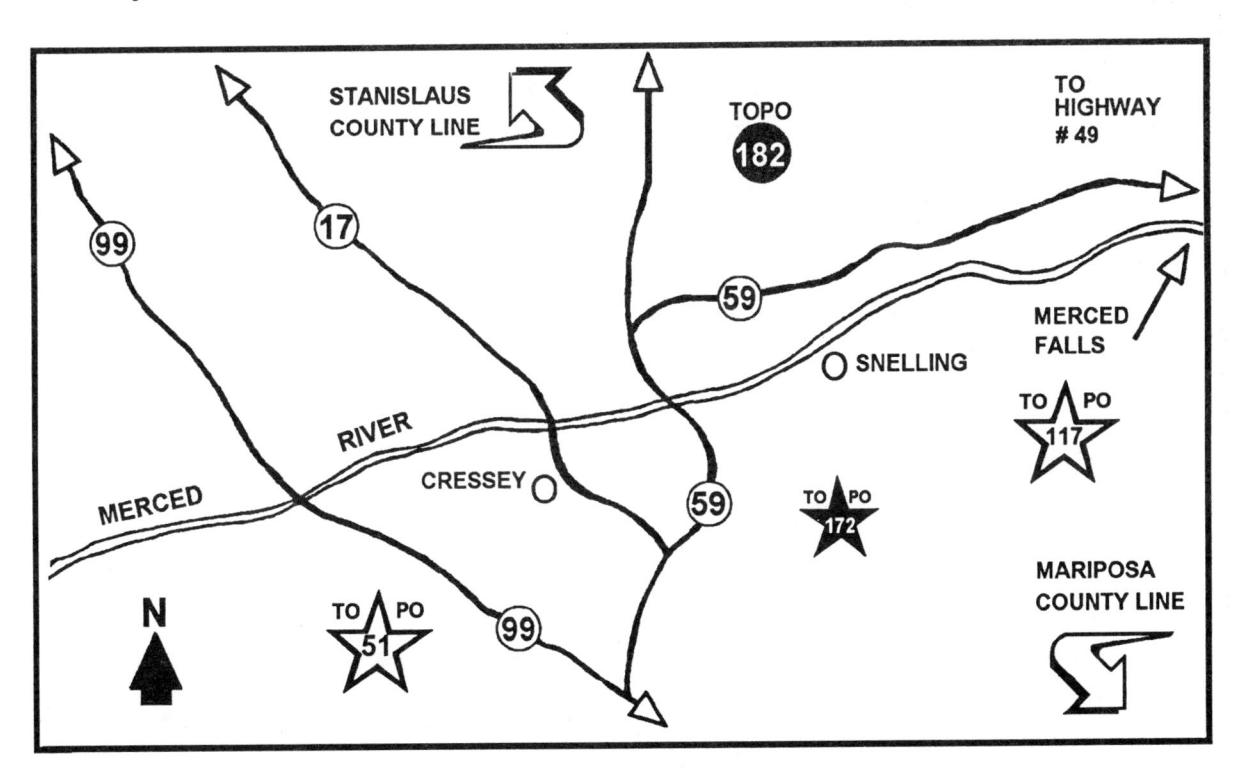

PIRU CREEK AND TRIBUTARIES, VENTURA AND LOS ANGELES COUNTIES. From the Santa Clara River upstream is *Class - A.*

TEXAS CANYON CREEK, LOS ANGELES COUNTY, the main stem is *Class H* open to dredging the year around.

SANTA CLARA RIVER AND ITS TRIBUTARIES, LOS ANGELES AND VENTURA COUNTIES are *Class -A, closed waters.*

Notice of intent can be filed at; Mt. Pinos Ranger Station, Star Rt. Box 400, Frazier Pk., CA. 93240. 805-245-3721.

The bureaucracy, in the not too distant past, did allow dredging the year around in the Piru Creek (excluding its tributaries) from the north end of Pyramid Reservoir, upstream to the confluence of Lockwood Creek.. check this out, as the rule makers may have retained this dredging as an active opportunity.

The placer deposits are adjacent to the upper end of Piru Creek, mainly in the vicinity of its junction with Lockwood Creek and eastward in the Gold Hill Area. **Gold particles** have been discovered from the more recent water-runs and from those ancient terrace deposits and channels. The back-country hosts a number of north-running gold quartz veins that contribute to the gold particles found in the water-runs from leaching and from percolating riparian effects.

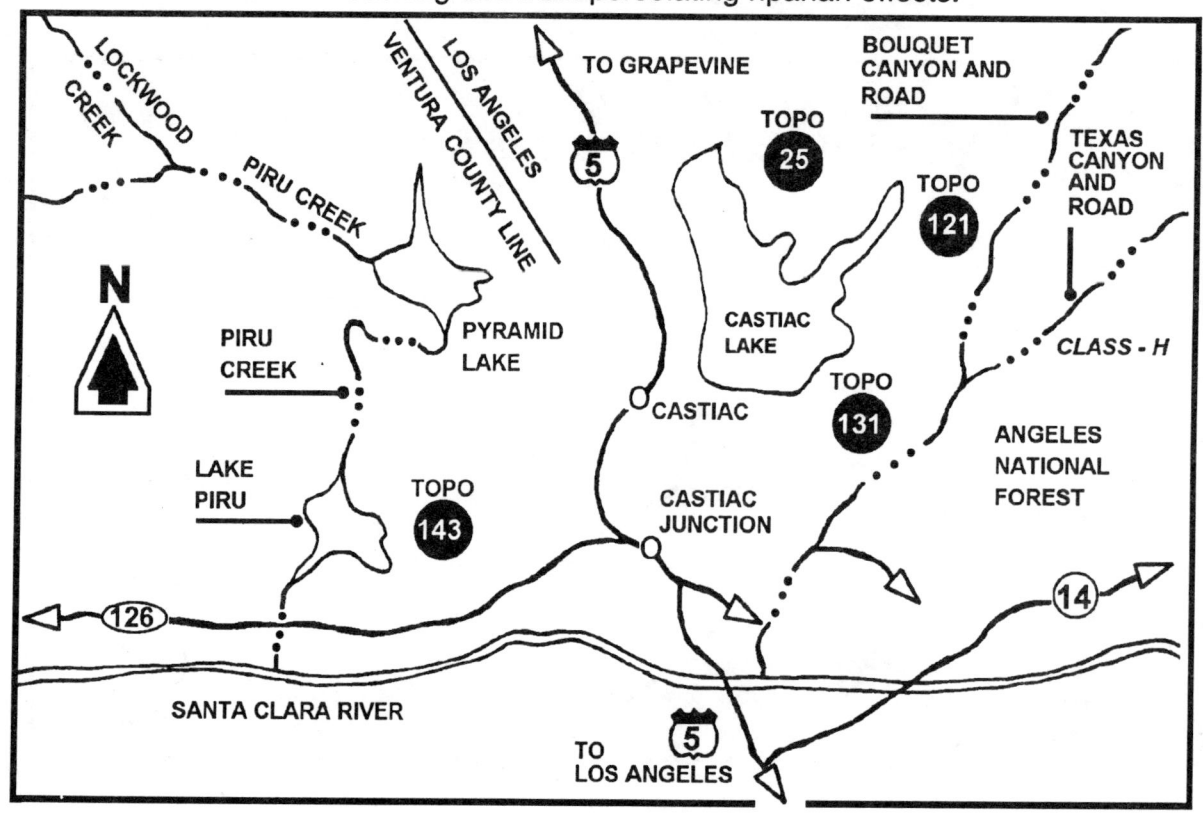

SAUGUS--NEWHALL, SAN GABRIEL MOUNTAINS. Los Angeles County is *Class - H.* Angeles National Forest. Check with local Ranger Station for area of interest for class openings in dredging, gold panning and dry washing activities. This gold district has many scattered housing developments, private mining properties with USFS Mineral Withdrawal Areas and scattered environmental sensitive areas. For workable Notice of Intent check with the District Ranger Station at Bouquet Canyon, or with the USFS Angeles National Forest, 701 North Santa Anita Ave., Arcadia, CA., 91006. 818-577-0050.

Placer gold in varying amounts and grades can be found in almost every stream that drains the San Gabriel Mountains. Tributaries adjacent to gravel Terraces, ancient stream-bed channels with scattered and hidden basins holding placer gold in their gravels. The entire Sugus-Newhall area has had a placer gold production history in the past. Most of this placer gold came from up-hill quartz veins, stringers in ancient rock and surface seated cemented gravel of Tertiary Age. Percolation from winter off-shore storms send the placer gold down into the feeder creeks and the tributaries of the Santa Clara River basin where they settle into locked-in depressions. Week-end prospectors work their way around the many gravel benches and Terraces that are affected by the seasonal flooding and gravel moving forces. There are several sizable areas in the USFS proper which are open to gold panning, gold sniping, dry-washing and placer mining operations. Check in with the USFS Office and review their Forestry Maps for areas of interest; Santa Felicia Canyon and its tributaries, Bouquet Canyon and its tributaries, Haskell, Dry Vasquez, Texas Canyon and its feeder creeks, Charlie, Tapia, Castaic, Bear, Pole, Mill Canyons and near the junction of most feeder-runs where gold placer can be located. Gold sniping is the best activity.

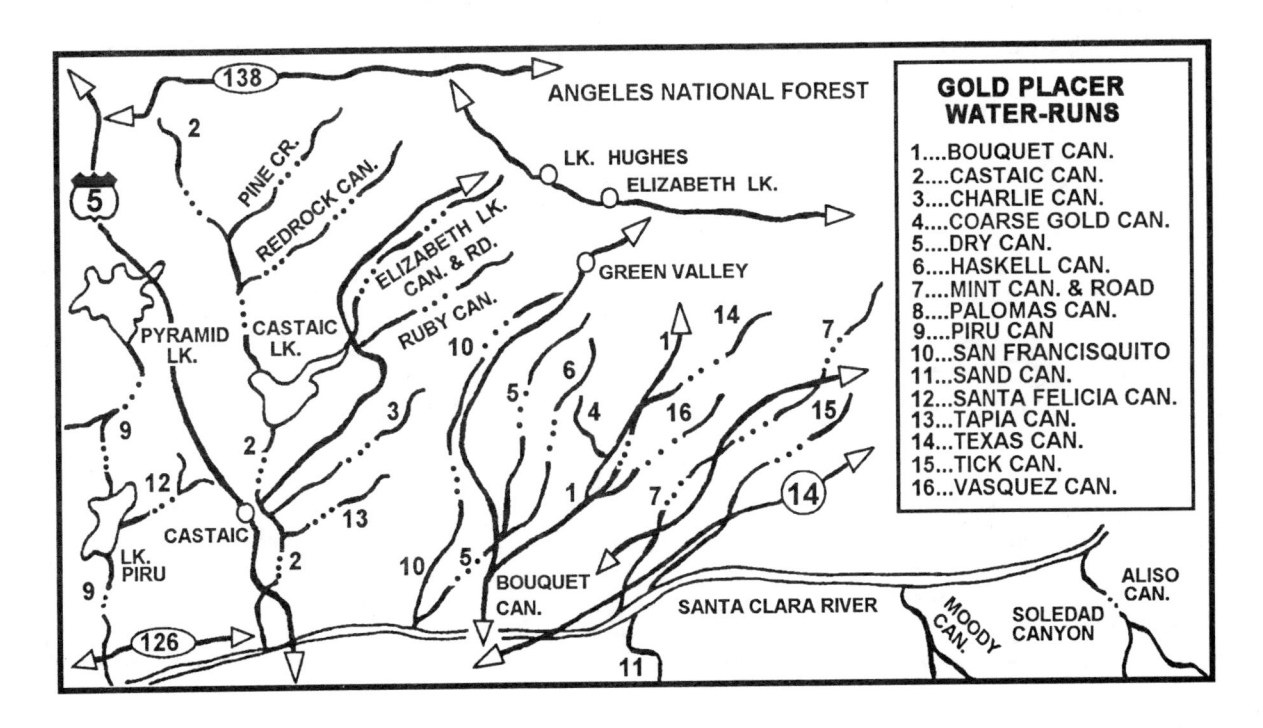

GOLD PLACER WATER-RUNS

1....BOUQUET CAN.
2....CASTAIC CAN.
3....CHARLIE CAN.
4....COARSE GOLD CAN.
5....DRY CAN.
6....HASKELL CAN.
7....MINT CAN. & ROAD
8....PALOMAS CAN.
9....PIRU CAN
10...SAN FRANCISQUITO
11...SAND CAN.
12...SANTA FELICIA CAN.
13...TAPIA CAN.
14...TEXAS CAN.
15...TICK CAN.
16...VASQUEZ CAN.

EAST FORK OF THE SAN GABRIEL RIVER, LOS ANGELES COUNTY.

Use USGS Topos; **Glendora, Crystal Lake and Mount San Antonia.** For location use Thomas Street Guide, California Highway Map and the Angeles National Forestry Maps. For information write to; Mt. Baldy Ranger Station, 110 North Wabash Ave., Glendora, CA., 91740.   818-335-1251. Rules and regulations change from time to time due to an old Congressional rule.

Exit from I-210 freeway off-ramp onto Highway # 39 , travel north thrue Azusa to the East Fork Road and follow this along the river to roads end. (Topo location T2N R8W). The river area is open to sluicing, dredging and gold panning. Other areas are open for sluicing and gold panning, while other areas are open to gold panning only. Area is open all the year around when allowed. Best to check with the area Forest Ranger Office.

Torrential seasonal rains create raging, forceful waters that rearrange old stream-beds and the present river channel while uncovering older bench gravel terraces. This abnormal freshet flushes down new gold particles and nuggets from higher elevation where the ridges have exposed gold veins sending these into the lower feeder creeks.

Best gold in dry season can be located in older river-beds, terraces and channels in the higher gravels of the San Gabriel Valley Corridor. Recovery of gold particles can be accomplished with dry-washer, or by hauling the gravel aggregate material down to the river's edge where it can be sluiced. Nugget sniping the river's main can reveal unclaimed gold particles wedged into the bedrock crevices and fissures. Check those old tailing piles and heaps left behind by the old miners, and, the hydraulic piles found further up on the river's higher bank. The back-country hosts old placer workings, abandoned lode mines and miners habitats.

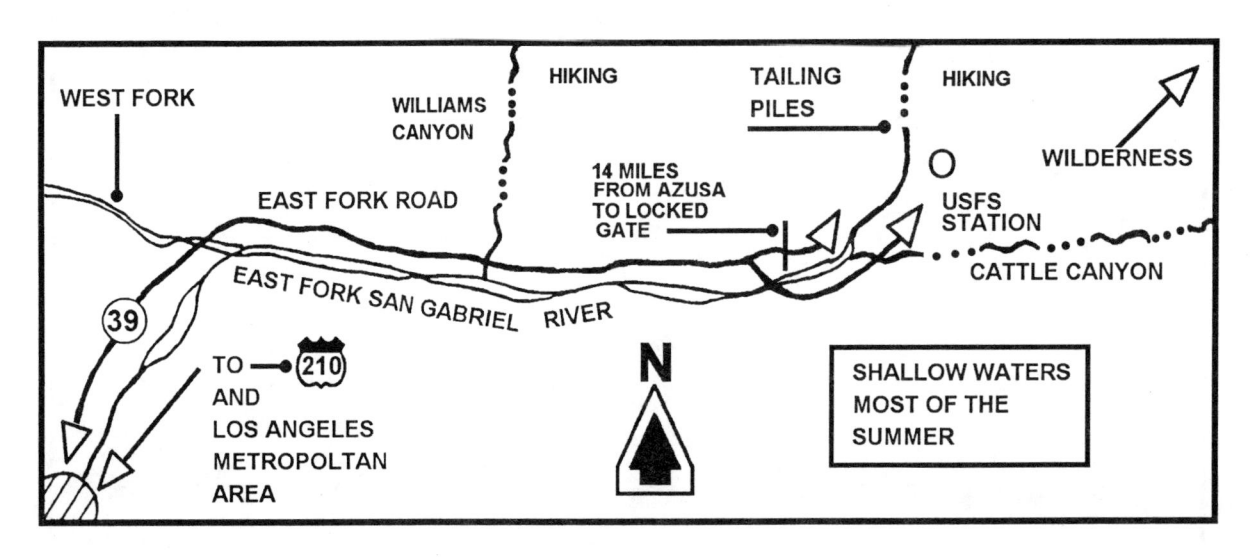

LA PANZA RANGE, (SAN LUIS OBISPO COUNTY), *is Class - A- no dredging permitted at any time.* Most of the season there isn't sufficient amount of flowing water to support any size suction drege operation. ***Gold panning, ground sluicing, sluice rocker and gold sniping with a metal detector is the norm for most of the working season.*** Existing gulches, canyons and ravines that slope drain from the eastern edge of the La Panza Range contain placer gold particles. Past history of placer production had been wrestled from the water-runs coursing thrue the gold-bearing cemented gravel of the Tertiary Age deposits gave impressive results. The Cretaceous gold-bearing deposits finger outwardly from the La Panza Range with present water-runs cutting thrue and re-depositing the gold particles into the lower riffles. Principle stream-courses which warrant consideration are; San Juan Creek, Navajo, McGinnis Creek, De La Guerra, South Fork Willow Canyon, Willow Canyon, Windmill, Cedar Creek, Carnaza Creek. Old Canyon, Placer Creek, Hay Canyon, Fraser Canyon and Beartrap Creek. For location use Topos; ***La Panza Ranch, Branch Mountain.*** The water courses and their dry-beds of the north and northwest end of the La Panza Range that have possible gold-bearing gravel are; Shell Creek, Fernandez Creek, Indian, Camatta Creek, Yaro Creek, Wilson Canyon, Alamo Creek, Huerhuero Creek, Bean Canyon, Toto Creek and Pozo Creek. For location use Topos; ***Chimney Canyon, Creston, and Pozo Summit.*** The tributaries and feeder creeks of the main-stem water-runs have cut thrue the rich gold-bearing conglomerates of the Tertiary Age gravel and have carried the gold occurrences downstream to drop into the lower riffle basins. Gold snip the old tailing heaps 'n piles. The southern end of the La Panza Range is off-limits to gold prospecting because of the ***Machesna Mountain Wilderness Area.*** Use Topo, County, California Highway and Los Padres National Forestry Maps for location and access roads. Be alert to private property and old mining claims. Best to check with local County Recorders Office for land rights.

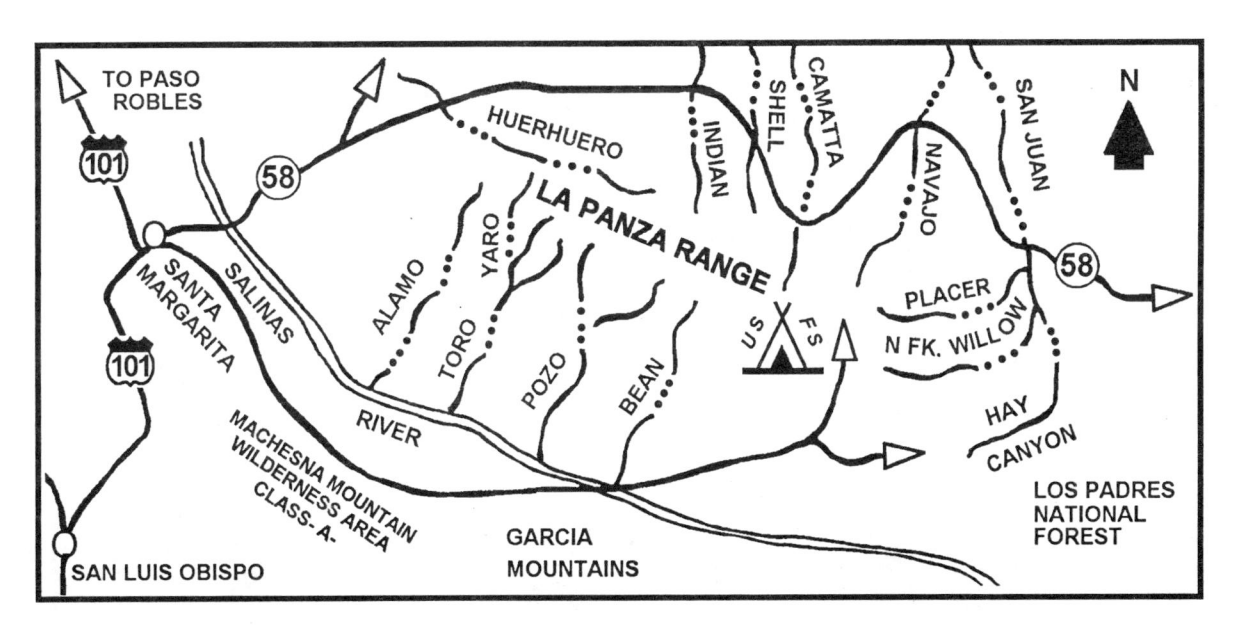

SANTA ANA MOUNTAINS, ORANGE AND RIVERSIDE COUNTIES. *Class H open to dredging the year around except where noted.* The mountain range consists of steep slopes and canyons buried beneath thick 'n dense chaparral growth making gold searching difficult at best. Many old mines have caved in and lay covered beneath the overgrowth making cross-country hiking, prospecting 'n searching a dangerous activity, and is only for the more adventurous person. The southern portion of the Santa Ana Mountains are sectioned off into the San Mateo Canyon Wilderness Area and are off-limits to any type of prospecting or mineral mining. Before entering the mountains in search for gold, best to check with the Corona Ranger Station at; 1147 E, 6th. St., Corona, CA., 91720. Regulations change frequently. Santa Ana Mountains are host to the Cleveland National Forest-Trabuco District. It has a history of placer gold 'n lode mining. The total amount of gold production was considerably more than what had been poorly recorded, as the records indicate that the early miners had found a substantial amount of placer gold in every water-run and canyon. Many of the hidden gulches contain a varying degree of placer gold, especially downhill from known mineralized areas and at old abandoned lode mines. Placer deposits show up in pockets scattered thrue the streamcourse systems. Most stream-beds are dry during the summer months. Winter is the best time to find water for dredging and sluicing in the canyons. Gold sniping is the best activity the year around for best results. There could very well be hillside placers, bench 'n terrace deposits which have remained undiscovered and overlooked in the San Juan, Trabuco, Silverado, Santiago, Verdugo and Lucas Canyons, along with their feeder systems, where placer gold can be located. This is very wild and undeveloped terrain, be alert to poison ivy-oak, snakes and mountain lions.

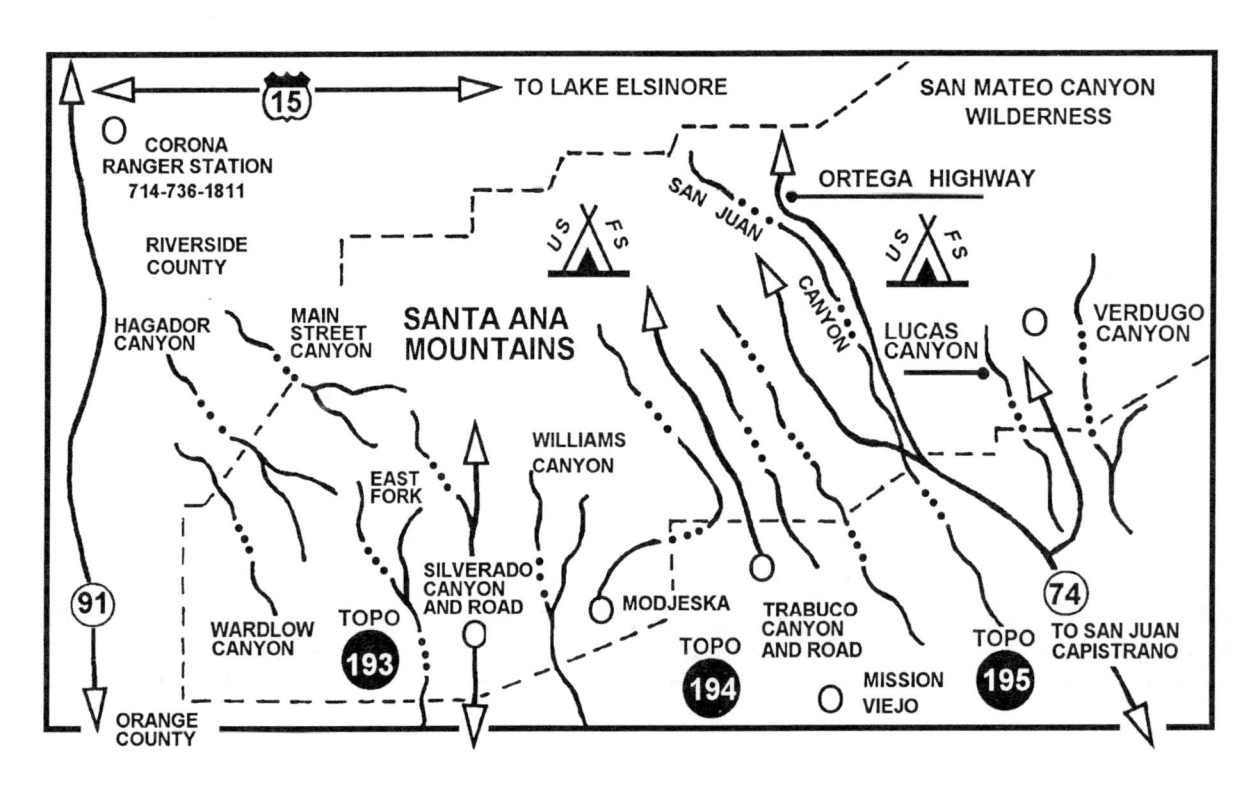

BOOKS by DELOS TOOLE

"Delos Toole's Where To Find Gold In CALIFORNIA" takes the gold searcher to the Klamath Mountains, Warner Mtns., High Sierras, American River, Feather River, Trinity River, San Gabriel Mtns. 'n River and lots of hand drawn MAPS with directions to the gold sites.

"Delos Toole's Where To Find Gold In OREGON" will provide the gold searcher with sites set aside by the US Forestry and the BLM depts. as Free-Use sites for dredging, panning and sluicing. MAPS with GPS directions to the placer gold sites and geological locations.

"Delos Toole's Where To Find ARIZONA'S Placer Gold" is chocked full of MAPS with directions to the gold sites for eleven counties. New untouched ground for metal detecting, dry washing, dredging and prospecting. Showing old mines and old placer regions on the MAPS for the searcher to prospect for rare earth and exotic minerals.

"Delos Toole's Where To Find YANKEE" Placer Gold leads the gold searcher to Maine, Vermont, Connecticut, New Hampshire and Massachusetts. MAPS on every page with directions to the gold sites. Roads to the water-runs are easily reached. Backcountry of Maine beckons the gold searcher. Gold particles in every stream in Maine, Vermont and New Hampshire. Connecticut and Massachusetts host their share of the gold.

"Delos Toole's Gold Nugget-Teering In NEVADA" Book Two, Is the companion book to "Delos Toole's Gold Nugget-Teering In NEVADA" Book One. Each book brings to the gold nugget searcher's need for Information with MAPS and directions to the placer mining sites. NEVADA is a highly mineralized state with undiscovered rare-earth metal deposits. Each MAP shows where mining has been active, and where the old timers mined before are good locations to search for gold nuggets. With each passing thunder storm racking the landscape and stirring the matrix into new riffle locations and laying those gold nuggets where the searcher can discover their gravity settlings.

Retail price for each book: $22.95 plus US Priority shipping $4.00.
Send check for $26.95 to: Delos Toole Gold Books, 5564 Lloyd CT SE Salem, Oregon 97301
Log onto my web page to order my books: Web Page http://www.delostooleauthor.com
E-mail delostoole@earthlink.net

CHEMICAL ANALYSIS TESTING KIT

The Kit consists of a variety of test equipment, chemicals, regents, test tubes and tools, chemical solutions to be mixed at home for making all of the tests for the 48 elements by numbers.

One book with a short course in exotic mineral identification with the use of the Chemical Analysis Testing Method.

One book with tests for Quick Quantitative and Qualitative Chemical Analysis step-by-step numbering procedure and system method.

One instructional book with tests in Beryllium and the rare earth minerals.

Inquires will not be acknowledged unless accompanied by an *SASE.* Send inquiries to; DELOS TOOLE   P.O. Box 3353, Landers, CA. 92285.

SOUTHERN CALIFORNIA. MOJAVE PLATEAU MINING

DISTRICTS. A wide area for placer gold, dry washing for gold occurrences and searching for old abandoned lode mines. Southern California gold mining districts are host to hundreds of old mines with debris dumps holding vast amounts of noble metals, rare earth minerals and exotic metals. Old miners of the 1850's abandoned the cracked rock because the metallic ore hadn't shown visible gold. In the course of moving tons of rock, the old miner uncovered and left behind fortunes in commercial ores. These lay as they were discarded just waiting for today's gold searcher, prospector and metal detectorist.

ALVORD district is in the central San Bernardino County, northeast of Daggett. Lode mines produced gold. Good placer dry washing area. **ARICA** is located in the Arica Mountains, in the northeast area of the Riverside County. Abundant of sulfides with gold-quartz in granite and schist. **ARROWHEAD** district is in the southeast area of San Bernardino County in the Providence Mountains, hosting high-grade pockets of free-gold occurrences.

BENDIGO district is located in the northeast section of Riverside County hosting deposits of gold, silver, copper and manganese. Dry wash gravels in the washes. **CARGO MUCHACHO** is located in the southeast section of Imperial County and east of El Centro with lode and placer deposits west of the Cargo Muchacho Mountain Range. **CHOCOLATE** district resides in the confined area of the eastern Imperial County and east of Glamis. At the edge of the southern and western foothills, placer deposits flank the Mountain ranges.

CLARK is hosted by San Bernardino County northeast of Baker. Gold-bearing deposits, rare earth metals and exotic ore deposits. **CHUCKWALLA** district is host to the Chuckwalla Mountain Range in eastern Riverside County, south of Desert Center with high-grade gold placer deposits. **COOLGARDIE** is a dry placer district in western San Bernardino County northwest of Barstow with gold residue created by several east and northeast veins.

DALE district borders San Bernardino and Riverside County with high-grade pockets of gold-silver in deeply seated veins. **DOS PALMAS** district is host to the Orocopia Mountains. Northeast of the Salton Sea area located on a serious faulted section with a sheared zone of gold-quartz. **EAGLE MOUNTAIN** district is in eastern Riverside County west of Highway # 78. Faulted fissures of lead and copper, gold-silver in dry placers.

GOLD REEF district is host to the Clipper Mountains in east central San Bernardino County and northwest of Essex, south of Highway # 40 (dry placers). **GOLDSTONE** district, northwest of San Bernardino County (Military area-closed). **GRAPEVINE** west of San Bernardino County with shallow dry placers.

HACKBERRY MOUNTAIN is northeast of Mitchell Caverns in San Bernardino County. Notable for rare earth, noble metals and placer gold gravels.
HALLORAN SPRINGS is northeast in San Bernardino County and south of Shadow Mountains, south of Highway US # 15. Placer gravel and gold stringers.
HART district is northeast in San Bernardino County in the eastern foothills of New York Mountains. Gold, rare earth metals, bismuth and placer gravel deposits.

IBEX district is in eastern San Bernardino County, north of Needles and Highway 40. Dead Mountain, Piute Mountain and Piute Valley with free-gold sulfides.
IVANPAH district is in the Ivanpah and Mescal Ranges with notable deposits of rare earth and basic metals, exotic minerals with placer gravel deposits.
MOJAVE-ROSAMOND district in southeastern Kern County, south of the town of Mojave where five brooding buttes influenced the area with gold occurrences.
MULE MOUNTAIN district is south of US Highway # 10 and west of Highway # 78. Native gold. pyrite and chalcopyrite occur in quartz veins and granite rock.
OLD DAD district hosts the Old Dad Mountains in northeast San Bernardino County. Shallow deposits of native gold and sulfides with placer gravels.
OLD WOMAN MOUNTAINS is in eastern San Bernardino County. Gold and abundant sulfides in quartz veins hosted by granite rock in shallow depths.

ORD district in west-central San Bernardino County, southeast of Barstow in the Ord and Newberry Mountain range straddled by Highway # 247. Lode mines.
ORO GRANDE district west-central in San Bernardino County and north of Victorville and bread 'n buttered by US Highway 395 and US 15. Hills host oxidized ore. **PICACHO** is in southeastern Imperial County 20 miles north of US Highway # 8 and west of State Highway 34. Dry washing placers. Check old mine ore dumps.

POTHOLE district in southeastern Imperial County and east of El Centro. Dry desert placers. Ravines with shallow pockets. Gold sniping area.
RAND or RANDSBURG district lies on the Kern-San Bernardino County line. Surface pockets of high-grade ore. Dry placer gold gravels.
SHADOW MOUNT district northeast in San Bernardino County and north of US Highway # 15 and west of Highway # 127. Check the old mine ore dumps.

STEDMAN district is in south-central San Bernardino County with ore-bearing zones in quartz monzonite and rhyolite. Fine examples in free-gold and rare earth nobles.
TROJAN hosts the Providence Mountains and Mitchell Caverns west of Highway no. US 40. Dry washing placers. Gold-silver-lead abandoned mines.

VANDERBILT district is in northeastern San Bernardino County within the New York Mountain Range, north of Goffs and east of Kelso. Dry washing placers.
WHIPPLE MOUNTAINS is southeast in San Bernardino County. Borders east of US Highway # 95 and west of the Colorado River. Look for old mine ore dumps.

IMPERIAL COUNTY GOLD....*THE POTHOLES*....PART ONE.

All of the land surrounding the POTHOLES is withdrawn from mineral entry. The area is Open To The Public as a Free-Use Recreational Gold Mining Area so long as there is no surface damage created during dry-washing operations. Use the Topographic Quadrangle map; *Bard, at location T15S R23E Sec. 25.*

The POTHOLES placer gold is located about twelve miles northeast of Yuma, Arizona, on both the California and the Arizona side of the Colorado River. The area has had a turbulent history as the result of incessant pursuit for its very rich placer gold and lode mines since the irascible and greedy Spaniards first appeared on the scene in 1779.

Gold nuggets located at the POTHOLES and at the Laguna Dam Placers are of the jagged, coarse and irregular shaped size. This area was formed by faulting of mineralized and shear zones in schist and granitic gneiss where the lower Colorado River embraces the land north of Yuma, Arizona.

The faulting of the bedrock created huge basins, deeply cupped depressions, humungus shaped cracks and fissures in the land. Later in time, land movements had scraped the rich gold veins of the northern lands into holding patterns. Still later the pre-Cretaceous forces pushed the gold filled glacial aggregate to the south, filling those cavity POTHOLES with rich gold concentrates.

Gold is found in the area at the mouth of most of the gulches and on the slopes of the Mountains where in the past have drained the mineralized zones into the Colorado River. Most of the placer gold was eventually concentrated at the base of the foot-hills at the higher-levels above the Colorado River.

Gold sniping with a metal detector is the best means for which to locate the POTHOLES gold nuggets. Dry-washing is one more means for locating the gold occurrences with. The best method is to spread the raw-concentrates onto a plastic sheet and allow this to dry thoroughly before working thrue the dry-washer. With finished concentrates placed into the container for easy hauling to the nearest water source to be washed thrue a sluice-box set up, is the final solution.

The area hosts several Bureau Of Land Management (BLM) Long-Term Visitor Areas where a 14 day limit use is allowed to be used as a base camp while prospecting the area and searching for the gold nuggets. For further information write to; BLM Office, 333 South Waterman Ave., El Centro, CA 92234.
619-353-5842.

POTHOLES. From the town of Winterhaven, located on US-8, travel north on S-24 past the Laguna Campground. Follow the eastern side of the All American Canal north until a gate is reached on the left side, cross over the canal at this point. Make a extreme left turn after passing thrue the gate and follow the west side of the canal south for a short distance until a side road is reached on the right side of the road. Turn right at this point, westerly, and travel a few hundred yards to the fork in the road. The left-hand fork takes the gold searcher to the POTHOLES. The right-hand fork leads off toward the LTVA free-campsites managed by the BLM. *Summer heat is harsh. Plan the trip in the fall, winter or springtime. Carry extra water and be prepared for emergencies when entering this area. Caution should be of the utmost concern. The LTVA is a favorite meeting place for the seasonal SNOW-BIRDS.*

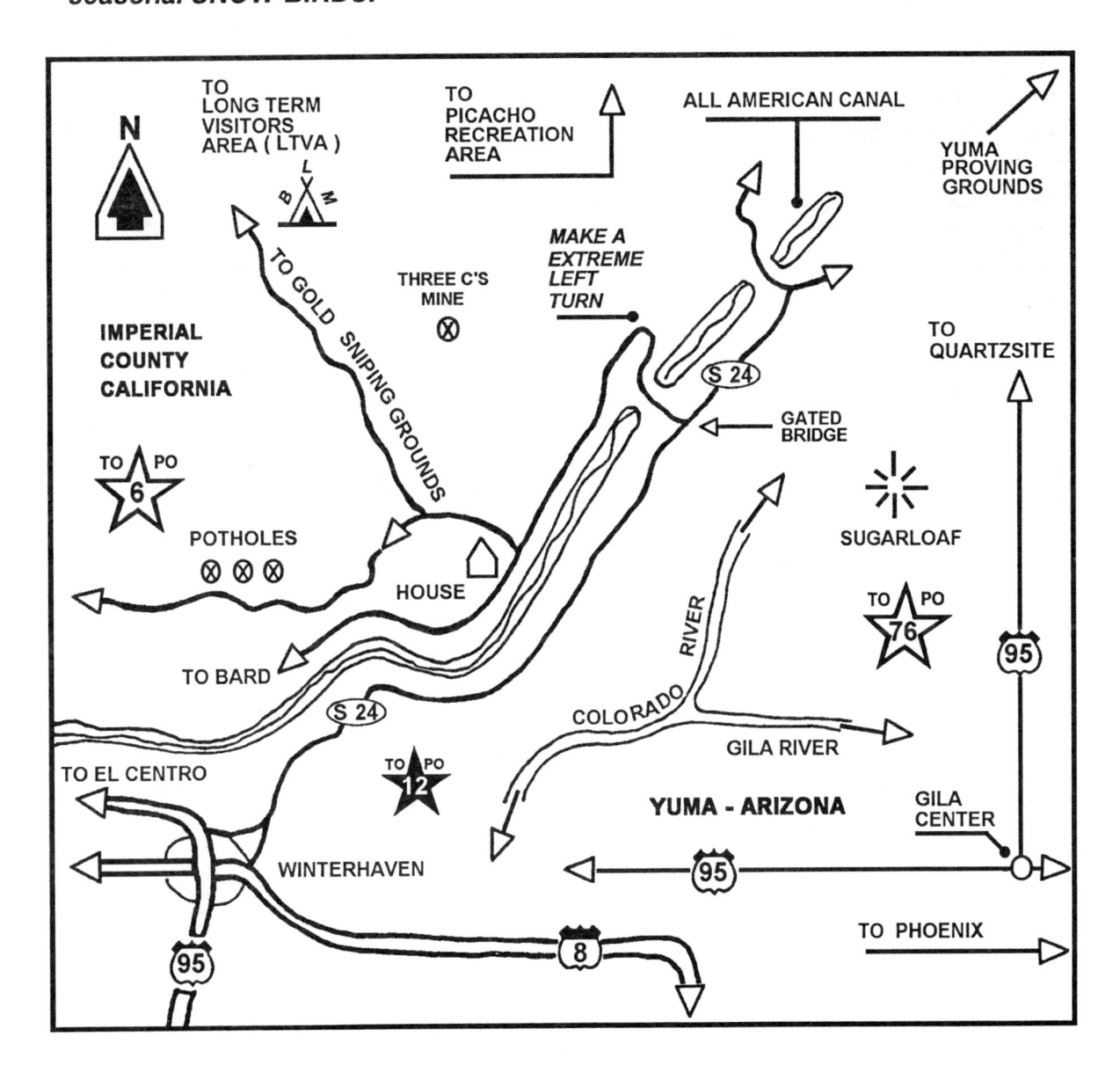

IMPERIAL COUNTY GOLD. CHOCOLATE MOUNTAINS PART TWO.

Use Topograghic Quadrangles, **Gables Wash, Mt. Barrow, Quartz Peak and Buzzard Peak for location.** The northeast, eastern and southeastern perimeter of the military aerial gunnery range of the Chocolate Mountains provides from its drainage system, gold particles caught up in the lower folds of its desert terrain that finger out from the parent elevation. Mineralized ledges at the higher elevations within the restricted area lay undeveloped and get raked by torrential thunderstorms that send the gold particles tumbling into the lower regions until the gravity pull slows down the gold movement in the cuts of the land. This attracts week-end gold snipers with their dry-washers and metal detectors throughout the seasonal year. Many mining claims in recent years have graced the gulches, ravines and canyons that finger out from the Chocolate Mountains. Most of these have since lasped back into the public domain and are open for search and exploration. Gold sniping, dry-washing and prospecting the gulches and canyons have produced for others a substantial degree of gold particles. For some, their mineral claim provides a "horn of plenty' to add to their income for each season. Be alert to existing mining claims and private property. Exercise caution during the summer season when the heat is life threatening. Take plenty of water, spare parts for the vehicle and personal savvy in survival know-how. Each year the desert claims many unsuspecting human lives from its harsh heat. Go prepared! Best time to enter the desert is in the fall, winter or early spring. Use caution and alertness to sudden thunderstorms that create flash flooding. Use a California Highway Map, BLM District Resource Map, Imperial County Map and the Automobile Club of Southern California Map (AAA).

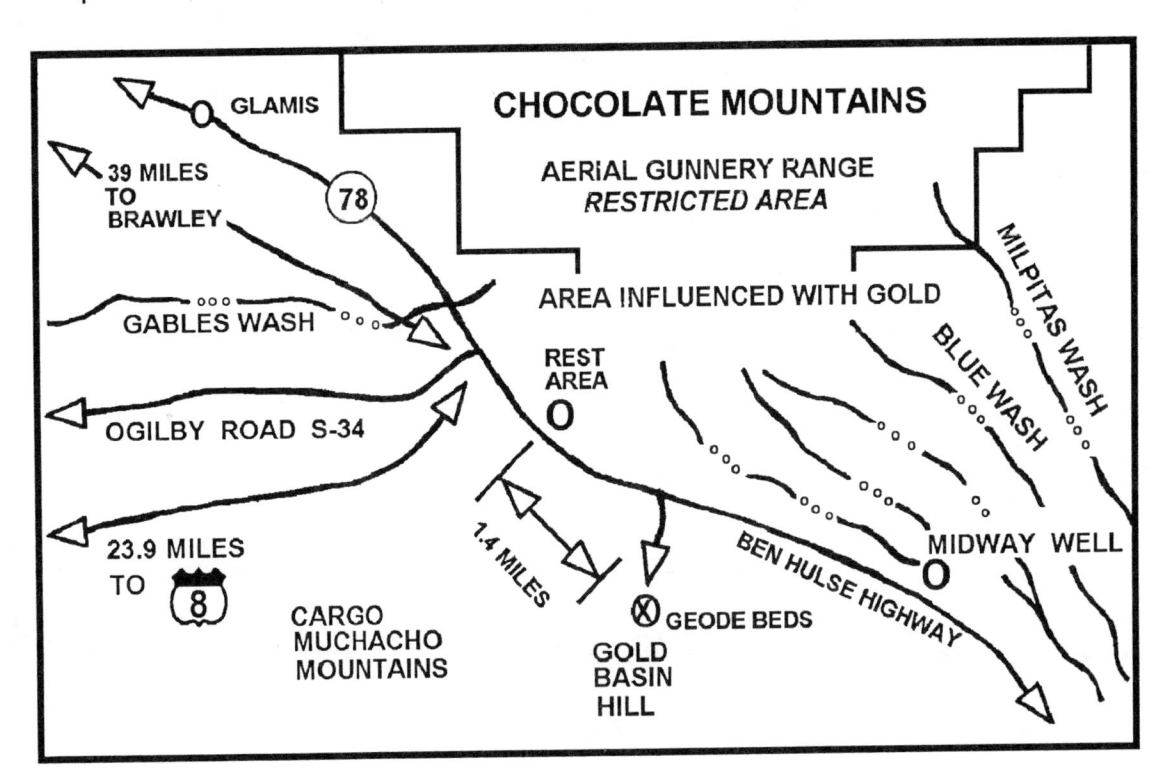

IMPERIAL COUNTY GOLD....MIDWAY WELL....PART THREE.

Use Topographic Quadrangle for location, *Quartz Peak, T11S R20E Sec. 16.* From the south direction leave I-8 traveling north on the Ogliby Road (S-34), for about 23.9-miles northeast arriving at State Highway # 78, turn right and travel about 9.7-miles northeast to Midway Well Road. Turn left onto this road and travel 0.4-miles west to a crossroads, turn left here and drive another 0.5-miles south to Midway Well Campground. From the north direction at Blythe, leave the I-10 and take the State Highway # 78 south for about 33 miles to the Midway Well Road and turn right onto this road. For a base camp, stay at Midway Well, a private campground with plenty of well water. No fee from April to October. The "snow birds" arrive from November to April and pay a small fee. To reach the dry-washing and gold sniping areas, drive across to the wash at the south end onto the BLM road (M151-sign), and make a right turn to the west. Stay with this road (M151) to the old mines and the gold sniping, dry-washing areas. Be alert to open and exposed stopes, mine shafts, fences and barriers. Beware of fragile timbers and rotting wood beams. "NO TRESPASSING" signs have been standing for the last forty years and at the present time these signs have no meaningful authority. The mines, lands and roads have reverted back to the public domain with the BLM managing the area and maintaining the roads. Gold snip and dry-wash the old mine dumps, gullies and ravines. Best chances of locating gold nuggets and placer deposits is in the remote areas of the adjoining canyons, ravines and gulches and feeder cuts in the land downhill from the old mines.

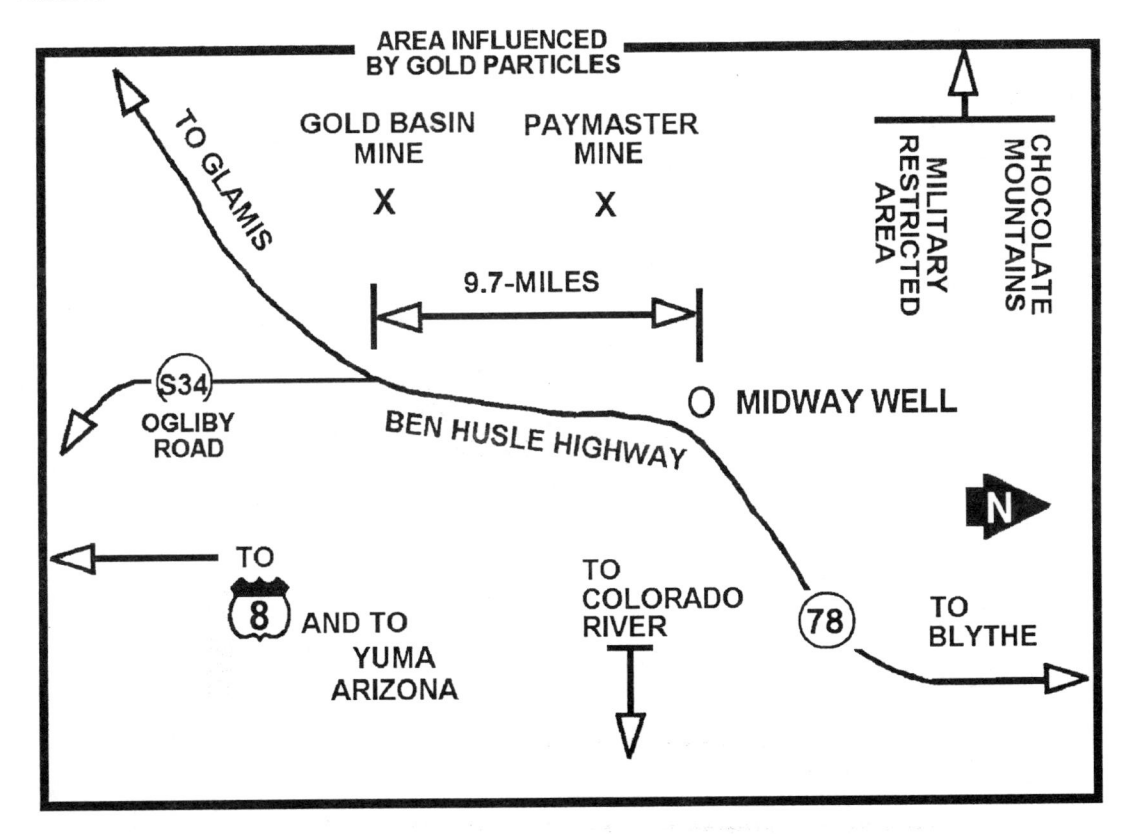

IMPERIAL COUNTY GOLD. CARGO MUCHACHO. PART FOUR.

Use Topographic Quadrangles, *Ogilby and Araz Wash for location.* Situated in the southeast corner of California with common borders with the Colorado River, the State of Arizona and with Mexico. Patience and determination with gold sniping for nuggets will prove out that the different washes, gulches and cuts in the land do contain a lot of gold particles that was left behind from the early 1800's boom-days. In particular, Jackson Gulch, Araz Wash, American Girl Wash and the Stinagree Gulch, all of which will expose the gold searcher to gold nuggets within their folds, or to other near-by gullies hosting gold nugget fields. Gold sniping is very time consuming but rewarding and should be entered into with a tenacious attitude to persevere in achieving the gold. The area has had a very active history in placer and lode mining. The old-timers didn't get all of the gold in the ancient cracks, crevices, fissures and cuts of the land. There is plenty of room in the Cargo Mountains for everyone to do their gold sniping, prospecting and dry-washing in the gravel benches. Gold sniping the downside of known lode mines for the ancient placer run-off is one of the best starts the gold searcher can make. Summer heat creates life threatening events and should be avoided during the seasonal period. Enter this land with plenty of spare water and automotive parts. Leave I-8 for about 4.1-miles on the Ogilby Road (north on S-34) just the other side of the RR tracks is where one meets up with the American Girl Mine Road. Travel this road for about 1.1-mile to a curve in the road to where a spacious gravel road appears. Travel this, following a row of utility poles in the southeastern direction which leads to a pile of old mine diggings that belonged to the old Cargo Muchacho Mine. Park your vehicle and foot-it the rest of the way to a near-by gully that fingers into Jackson Gulch in a easterly direction. During heavy rains this area gets its share of the gold particles.

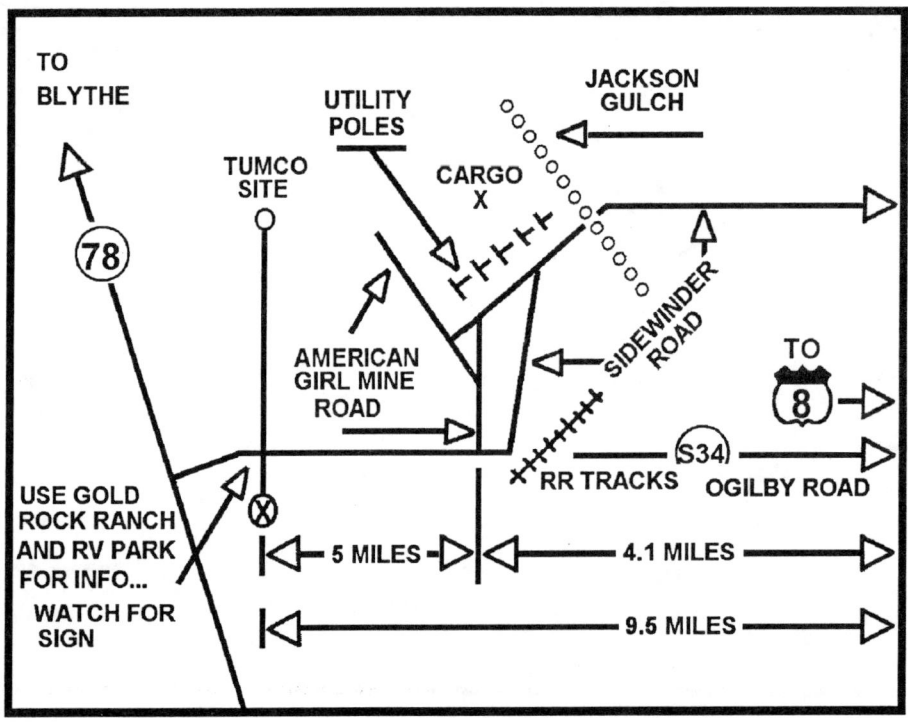

133

JULIAN-BANNER DISTRICT GOLD BELT. SAN DIEGO COUNTY.
CLEVELAND NATIONAL FOREST. Use Topographic Quadrangles for location; *Julian, Mesa Grande, Santa Ysabel and Escondido.*

Most creeks running in gulches, ravines and canyons flow only during the winter season. Thick heavy brush in most gulches makes prospecting hopeless. The Julian District has a great deal of private land, mining claims, locked gates, developments and restricted areas. Check access status with the Descanso Ranger District, 3348 Alpine Blvd., Alpine, CA., 92507. 909-697-5200.

Best placer ground is near Julian and Wynola, with other sites downstream from Banner in scattered open small parcels at T13S R4E Sec. 10 thru 14, 23 and 24. East of the Ballena Hills at T13S R2E are gold-bearing ancient channels from 30 to 90 feet thick and over 2000 feet in width. These areas are intermittently being mined on the ridgecrests at T13S R1E Sec. 16, 17, 19 and 20. Look for open areas in-between canyons of Hatfield Creek at location T13S R2E Sec. 18 and Swartz Canyon T13S R2E Sec. 21. Banner Creek T13S R4E Sec. 2 and Coleman Creek T12 R3E Sec. 36, along with their feeder creeks, report good color to coarse flakes in gold. This area is popular with dredgers. There are many scattered open areas which have the possibility of rich undiscovered placers buried in streams, gulches and ravines with bench gravels, ancient Terraces of cemented gravels in pockets. There are many old mines scattered thrue the San Diego County proper with old ore debris dumps. Metal detecting the dumps for overlooked gold nuggets in many cases has been profitable. The old debris dumps may contain rare-earth metals, exotic elements and minerals that were un-recognized by the old time miners. Chemical Analysis Testing in step-by-step sequence will prove this out.

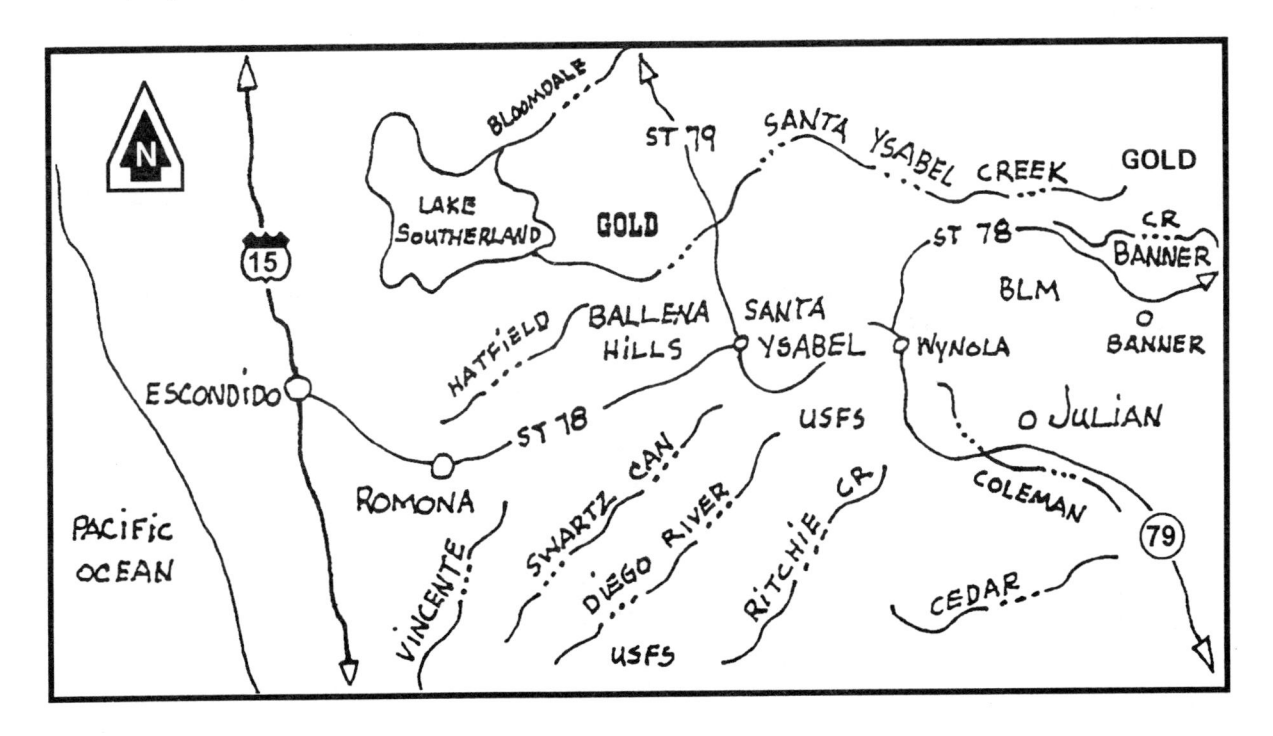

ORE DUMPS MAY HIDE MODERN DAY BONANZA IN RARE EARTH MINERALS, EXOTIC METALS.

by DELOS TOOLE

The rush of the 1990's interest in old ore dumps will be comparable in magnitude to the rush of the old 49ers who chased the gold nugget fields of the west. All through the 1800's, gold nuggets were scooped up from the surface of the land like one would pick strawberries today.

When the gold nuggets ceased to be easy pickings, the eye-balling prospectors turned their efforts from surface searching to digging holes, and soon discovered just below the surface of the land more of the gold nugget bonanza in glory holes, pot holes, shallow sinks and dried up water-runs.

Those eager prospectors stirred up the dirt and gravel like so many busy gophers, leaving behind discarded piles of rock containing unrecognized metals, unseen valuable elements and rare earth minerals

Time was when the old timer's eager quest for wealth was concentrated on only for gold. But world events created a demand for silver and this demand created more holes to be dug by the prospectors who threw away more seemingly worthless rock and dirt onto newly formed ore dumps.

Time marched on into the future and interest in new minerals appeared and created a demand for new test holes, and new ore dumps at the mine head.

Fueled by several wars, the world's appetite for more and different minerals seemed unsatiable and the hunt was on for tungsten, zinc, copper, lead as well as the dependable gold 'n silver.

This demand for new metals drew thousands of dreamy-eyed would-be prospectors out of America's depressed areas into the western part of the country to dig still more holes, sink shafts, carve tunnels and create new mines.

These new diggers sent the dirt flying in all directions, creating newer piles of dirt 'n rock and, unknowingly, creating high heaps of invisible wealth in rare earth materials, exotic metals and unseen valuable rare nobles, to lie in wait for future prospectors armed with new knowledge and technology.

The old-time prospectors, unaware of what treasures remained locked in the rocks they discarded, threw their split rock into the canyons below. That rock contained ores rich in cobalt, rare-platinum minerals and valuable bismuth.

Early prospectors, being "sight prospectors", passed over much valuable ore which gave no evidence as to signs in metallic form. This discarding moved tons of dirt 'n rock debris onto dumps, considered at that time as worthless material.

Fortunes in unseen minerals were left behind when the old mines were abandoned after the visible gold 'n silver, lead, zinc, copper and tungsten ran out. These old ore dumps have lain undisturbed for more than a hundred and fifty years, just waiting for today's knowledgeable prospector to come along with a field testing kit to chemically analyze its mineral content and to identify each individual metallic mineral, rare earth metal and mineralized element.

(continued on the next page)

The instruction kit that comes with the step-method of chemical analysis testing is easily understood. With varied alterations in heating tests, evaporation of the solutions showing colors in spectrum, step-additives in numbered chemicals and methods of filtration will give results and answer questions as to the unknown identity of the ore. It will also give an indication of its worth, quantity, percentages and will provide the prospector with reassurance that they have made a quick, intelligent assessment of the discovered old ore dump.

The chemical analysis testing is semi-quantitative, semi-qualitative and gives results of the ore sample tested in grades of very good, good, fair or poor.

Beryllium is a valuable rare metallic chemical element used to form strong, hard alloys with several metals, including copper and silver. This miracle metal, when used as an alloy, makes tools non-sparking, non-magnetic instruments, high-speed bearings and turns stainless steel into a "supermetal".

There are 48 very important rare minerals associated with beryllium and a simple test can determine what the sample host ore contains. Should beryllium or any of the companion rare earth minerals be located in low or high grade deposits, or in pockets of varying quantities near the earth's surface, it is possible to sell 100-pound bags of the material, giving the prospector quick cash without having to become involved with tons of ore by the carloads, as when dealing in lesser hard rock metals. The 18 beryllium minerals are the "poor man's road to easy street," along with the overlooked 66 minerals found in the same beryllium family formation uncovered using the easy step-by-step quick chemical analysis testing kit.

The following test illustrates how easy and simple this step-by-step analysis can be. Keep this explanation on file as it is an easy test for beryllium.

Begin by crushing to powder a piece of ore suspected of containing beryllium to make the Quinalizarin Solution which will be used throughout the test. NOTE; The testing kit which the author uses contains complete supplies for the beryllium test.

Place 60 drops of cold, distilled water into a clean test tube. Take sodium hydroxide, one-half the size of a green pea, and add to the test tube. Shake and agitate until it is dissolved. Into this solution add quinalizarin dye powder, an amount equal in size to a half-grain of rice. Mix and shake until this prepared quinalizarin solution has a purple color to it.

Prepare the mineral fusion by placing borax glass in the amount of one-half the size of a green pea onto a piece of chinaware or broken crockery. Make a small depression in the center of this and add a bit of powdered mineral sample from the old ore dump discovery equal to the size of a large grain of rice.

Next take a pellet of sodium hydroxide, half the size of a green pea, and place flat side down on the powdered sample. With a blowpipe, fuse this completely in a lamp flame. NOTE; While still hot, remove specimen sample with a knife blade and repeat the fusion over again. Once more while the specimen is still hot, remove it with the knife blade then crush to a powder in a porcelain mortar.

The next step is to place this crushed fusion onto a clean evaporating dish and add 20 drops of very cold distilled water using an eye dropper. Stir and mix to help dissolve the mixture. (Use this as a test solution to follow). (continued on the next page)

Take two small dishes to be used for separate tests. In one of the small dishes, place 2 drops of the test solution. In the other dish, place 2 drops of clear, cold distilled water; we will call this "blank" to be used for comparison.

To each of the dishes add two drops of the quinalizarin solution. NOTE; The blank will have a purple color to it. The dish containing the test solution will have a light blue color to it. If either beryllium or magnesium is found in the test solution, they will be easily distinguishable by the purple color found in the blank dish.

Should the test solution not produce a blue color, then the test is complete. Continue on with the next step-by-step of the quick chemical analysis testing. If the test does show the color blue, then proceed further testing to determine whether the sample of the crushed ore has beryllium or magnesium in it.

Place 8 drops of this test solution into a test tube along with 4 drops of the quinalizarin solution. Mix and shake well. If there is magnesium in the crushed ore sample, the test will turn the solution blue and after five minutes the solution will appear cloudy with tiny blue particle specks in the solution.

Place the test tube in the rack and allow it to set for 30 minutes until the tiny particles begin to settle to the bottom of the test tube as a dark blue precipitate; the solution itself will appear as a colorless solution.

If beryllium is in the crushed host ore sample, there will be a clear-like blue appearance to the solution. There will be no blue particles or dark blue precipitate show up in the bottom of the test tube as found in the previous test after setting for 45 minutes; the solution will now remain blue.

Somewhere out in the "boonies" lay thousands of abandoned prospect holes, old ore dumps and deserted mines where rocks were thrown aside onto the ore dump that could possibly contain a fortune in rare earth materials.

Maybe that helpful old prospector with his magnifying eyeglass, after looking at that piece of rock that was shown to him, called that black weighty junk as worthless and it most likely was the chief ore (columbite) of the element columbium which appears to be just common black iron at first glance; a missed treasure, to be sure.

Those black, brown or red nodules which so stubbornly stuck to the riffles in the sluice box and clung to the sides of the gold pan were possibly not magnetite or hematite, as the prospector thought. With the use of the chemical analysis testing kit those nodules would have shown up as cassoterite, the number one important ore of the element tin.

Perhaps the magnifying eyeglass the searcher was using fooled them once again as they eye-balled the red or brown granite rock speckled with black formations. With a few minutes of analysis testing, the black specks might have proven to be one of the many lost fortunes in rare minerals: tantalite, columbite, samarskite or cassiterite.

The curious doesn't have to know what one rock or element looks like, nor is it necessary to have an education in the sciences, or the ability to understand the mysteries of chemistry. Throw away the old fashioned idea fostered by the 1837 Dana practice of specific gravity, hardness tests, color and physical properties, and rely on chemical analysis testing.

TREASURES AND FORTUNES AWAIT THE SEARCHERS ON THE OLD RAILROAD RIGHT-AWAY AND ANCIENT RAIL BEDS.

by DELOS TOOLE

In its early years the Western United States had developed hundreds of mining districts, which in the interim had hosted thousands of mines that saw workers laboriously hacking at the insides of rugged mountains, tunneling as they went, and in so doing they produced heaps of dump material on the aprons at the mine head entrance.

In the beginning, many railroad beds making up the right-away were basically built with imported ballast aggregate. As the demand for shipping the ore increased, so did the need for bedding material to make repairs with for the existing railroad beds and for extending its line of operation.

The railroad builders quickly became aware of the great heaps of available dump material at the mine head from the tunneling efforts of the miners. This excessive waste ore rock was offered to the railroad builders free of charge if they would just remove it from the mine dump area and the mine entrance.

This new turn of economical events created a surge of competitors to service those developing mines in the newly formed mining districts and had attracted hundreds of railroad entrepreneurs to build road beds with the waste ore rock. Those railroad builders sent masses of polyglot workmen onto the mine heaps with their hand shovels, wheeled barrows, buckets, and dust raising horsemen urging their steeds to pull the scoop-haulers loaded with the mine dump material out to the ends of those extensions of the raised railroad beds mile upon mile.

Railroad spurs, branch lines fanned out connecting with the mainlines and before they were hastily established, the railroad bed was built and raised to span shallow ravines, gullies and low-lying depressions in the land. The bed material that the railroad builders had used came in the form of hacked rock of a spatial quality in unsystematic serrated edges and was better suited to the tonnage's hauled over the railroad beds.

In the long vacant deserted flat areas where there was no evidence of mineralization, the railroad cars hauled the loose rocky material from the nearest mine dump and extended the railroad bed for miles with this ore material. Wherever a wash-out from flash floods occurred in the railroad bed, the railroad cars and human laborers transported the dump material to the break site, making repairs to the railroad bed with the discarded waste material from the nearest mine apron or from previously stored heaps at the nearest siding.

Most all of the secondary, sub-railroad lines and branch lines were hastily built to service the demands of the small mining towns tucked away in many of the neighboring canyons where scattered tunneling existed in lonely placements. Shortline and feeder railroads were created on a shoe-string operation just for the moment, and ahead of other railroad competition close on their heels. The builders relied on easy accessible mined waste ore rock with which they built the road bed with.

Crushed aggregate, mined cinders, slag and slate were very expensive to process, as man-power was scarce and labor problems plagued the railroad builders. It was cheaper and easier to obtain the already bucketed mined waste ore rock from the mine head, especially for the money boys with quick get-rich schemes in stock 'n bonds which smacked of fraudulent promotions, and who ran very few trains over the new routes once the railroad beds had been built.

(continued on the next page).

As years came and went, rails that were once lain down were now torn up and were abandoned as the different needs dictated. Hundreds of abandoned railroad beds continue to exist today in remote mining regions, some alongside of modern highways obscured from view by vegetation, others are barren of rails but continue to scoot alongside of bleak escarpments. Several thousands of miles of these old railroad beds point in the direction where abandoned mines once flourished and are in evidence as to where the old mining districts existed. Other abandoned rails and railroad beds are found skirting in and out of existing and non-existing ghost town sites like so many phantoms, leaving a pointing finger towards the possible location of these sites.

The waste ore rock making up those thousands of miles in abandoned mining railroad beds can be analyzed by chemical testing where positive results will indicate that valuable modern day rare earth metals and their components exist. The old railroad beds will point to the nearest mine or mining district from which the waste ore came from and to where the searcher can locate the mine source that holds the existing remaining valuable rare earth ore dump.

Waste ore rock used for the railroad beds in itself would be considerably valuable should it contain one or several of the forty elements hosting marketable metals, such as the primary host ores of beryllium, beryl and bertrandite, that when properly processed in the metallic form are used for industrial and defense applications.

With metal detector in hand, the searcher should keep in mind that all railroad beds at one time or another required a great number of workmen, railroad crews, horse flesh and equipment to accomplish the work with. Work sections, work camps were established about every two miles apart, where stations were built about every five miles apart and terminals were established about every fifteen miles along the railroad bed and its trunkline.

Human activity created some tent 'n wooden communities, where with time these have become ghost sites in memory and are nothing more than vague indications that there once was much active activity here. Where people favored a certain area and location with their living impressions, they left behind many lost items and discarded everyday equipment which are considered antiques and as little artifact treasures in railroad memorabilia.

Scattered about are remnants of their living needs in old milk cans, sad irons, bottle dumps, lanterns, cooking utensils and a host of ancient miscellaneous. On each side of the railroad bed and it's right-away at arms throw, can be found pumpkin seed and embossed whiskey bottles scattered about in the landscape right where they landed when thrown out of the train windows or from the flat cars that the workmen rode to work on. These objects lay shining and sparkling in the sunlight and are seen as broken pieces of broken glass shimmering in their cobalt blue and purple hues laying amongst the rock and heavy vegetation.

While with one's recently acquired chemical analysis testing kit in hand, be alert to storage heaps in and near a likely railroad station, side tracks, section site or at a work camp site where waste ore rock was dumped into piles as a reserve for possible future railroad bed repairs from flash flood wash outs. Those heaps could contain a fortune in forty elements, rare exotic and rare earth minerals. Analysis testing of the heaped ore can be accomplished in just a few minutes of the searchers time and could reveal rich gold 'n silver telluride's that could not have been seen in the metallic form by the old eye-balling searching prospector and miner.

(continued on the next page).

Scattered about by hastily grouped repair centers strung out along many of the railroad beds are hand tools, locomotive and train parts, blacksmith's repair equipment and supply depots, all camouflaged by eroding wind, sand and time to test today's detectorist's ability with their balanced metal detector.

A serious antique and bottle collector would be wise to bring along to those sites a quarter inch mesh, hand sifting rack, for when their metal detector sings out at a strike. Gold coins are wherever a person finds 'em and bottle pits are ganged close to the ghost sites. Cached holes are near the old out-buildings, post-hole banks are where posts once ringed the horse corrals, chicken coops, cattle holding pens and abandoned Chinese garden plots.

Old railroad beds offer an easier access to the old ore dump values than those hard-to-get-to old mines would. For the less adventurous, or for the ill-equipped searcher without a 4X4 gravel grabber, the old railroad beds point to up-hill fortunes and give up ancient secrets when chemical analysis testing methods are carefully used in detecting rare earth minerals amongst the waste rock of the old railroad beds.

Most test kits used will produce positive results as an easy and simple method to unravel the mysteries, a child of the age of ten can understand the system and make it work. Instructions are easily followed, step-by-step, number-by-number in sequence through forty elements to its conclusion. That which is not seen, quickly becomes visible through the use of analysis testing.

For the reader East of the Mississippi River, there are thousands of miles where railroads have been abandoned leaving the rails intact, and many more that have railroad beds vacant of steel rails. Getting around on the rails takes some ingenuity and mechanical ability, but it is well worth the try. One can build a three wheeled vehicle or a four wheeled bicycle that will span the rails where the rider can peddle their way into history.

Take a metal detector into those remote and seldom seen areas where old trails once crossed, old settlements existed, old forts and abandoned rail station buildings once existed along the old abandoned railroad beds, and these become a challenging search 'n detect mission.

The possibilities are endless, as very few detectorists are aware that those sites ever existed, or, of knowing how to get into those areas, only the railroad beds will take the reader into those ancient sites. The rail-vehicle will take the detectorist down back-ways, behind many of the obvious small towns, past old deserted farm buildings and through hidden 'n forgotten areas to seldom experienced adventures in bottle collecting, lost cache sites and a host of desired relics.

There are clubs, rail groups and railroad buffs who use the abandoned railroad beds with their make shift rail-vehicles, just for the lark of it. Many abandoned railroad beds are rail-less and are used for hiking, bicycling 'n camping and are used for many varieties in recreational pursuits.

By doing a little researching, the reader will come up with some dandy experiences in detecting along the old railroad proper and right away. Remember, that many of the abandoned railroad beds are private property. Researching will unveil which right away has been released back to the public for their use. Be cautious of those rails which still have railroad traffic on them. The responsibility lies with the individual to avoid private property.

(continued on the next page).

The following list of railroads have been abandoned sometime between the years of 1945 and 1955. The list does not necessarily indicate that the railroad beds have been built with waste ore rock, or made up of valuable rare earth ores, quite the contrary, these probably do not. In some cases the abandonment has been purchased either by the water and power companies, ranchers
and by other railroads operating in the area; careful researching by the reader will uncover those facts.

This list is given only as an indicator pointing to the back country where ancient railroad beds do exist. Researching, exploring and searching out from those points will give the reader a better understanding as to where they might begin their search from and for the waste ore rock of unlisted railroad beds. Researching the local library, historical societies, State Mine and Geology Departments, State and private railroad museums will reveal where the old mines, the old railroads and abandoned ancient railroad beds are located.

Arizona.............Maricopa to West Chandler..............Postom to Florence.
California..........Laws to Nevada State line................Ludlow to Nevada State line.
Colorado...........Antonito to New Mexico State line....Texas Creek to Westcliffe.
Idaho................Salmon to Gilmore...........................Leadore to State line.
Montana...........Armstead to State line......................Roy Junction to Roy.
Nevada.............Battle Mountain to Austin.................Palisades to Eureka.
New Mexico......Mount Dora to Farley........................San Ysido to Marion.
Oklahoma.........Page to Pine Valley..........................Frey to Oilton.
Oregon.............Newberg to Tillamook to Gate...........Lackamas to Swift.
South Dakota....Conde to Akaska..............................Gettysburg to Redfield.
Texas...............Snyder to Fluvanna..........................Seymour to Mineral Wells.
Washington......Curlew to Midway.............................Mount Vernon to Finney Creek.

Hands & Pans Free-Use

Casual Mining on the South Yuba River

Tahoe National Forest

Tahoe National Forest
Nevada City Ranger District
631 Coyote St. Nevada City CA 95959
530-265-4531

Topo; Washington.
Nevada County, California.

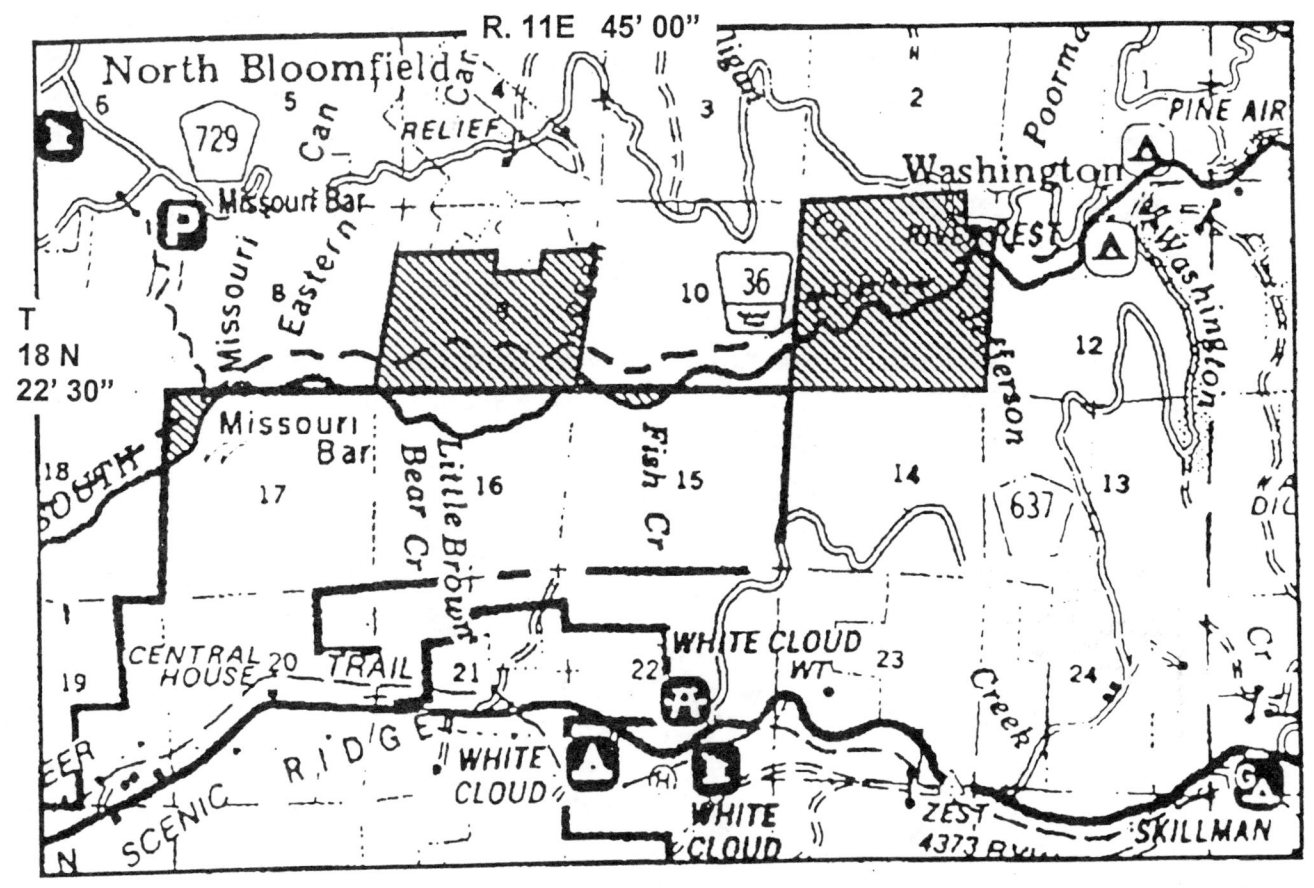

"Hands & Pans" Area: The shaded areas above are open for casual mining only. They are not open for mineral exploration or extraction under the 1872 Mining Law.

Directions to the South Yuba River Mining area

From Hwy. 20 east, approximately 14 miles to the Washington turn-off, turn left on the Washington Ridge road through the town of Washington; cross the South Yuba bridge, take first left, continue on this road until you come to a split in the road, (DO NOT TAKE GASTON RIDGE ROAD) stay to the left on Relief Hill Road. Continue for approximately 1 1/2 miles to the South Yuba trailhead. As you continue up the hill, looking down to your left, you will see the restrooms and parking area for the South Yuba trailhead.

THE FOLLOWING TEXT AND MAP IS AN EXAMPLE OF WHAT CAN BE EXPECTED IN THE BOOK BY DELOS TOOLE;
"Where To Find ARIZONA'S Placer Gold".

MULE GULCH - GOLD GULCH WASH PLACERS; Topographic Quadrangle Maps; Bisbee, Bisbee NE, Bisbee SE. Cochise County of Arizona. The accompanied map shows existing and estimated placer gravels indicated by stippling. This does not guarantee absolute placer existence, but rather it should be viewed as an indicator and as an aid for furthering placer gold explorations. The immediate area to Bisbee is heavily engaged in copper mining activities and should be avoided because of extensive private property. To the east at the breaks and foothills in the Mule Mountains are gulches, arroyos and canyons that have in the past been placer mined and continues to offer possibilities. Activity has been limited to the paleoplacers of Glance Creek by a few well-informed placer miners. Glance Creek and its many gulch tributaries were formed by Jurassic age mountains being eroded and deposited into formations of pediment mesas, fans of alluvials and gravel channels from ancient Jurassic and early Cretaceous age.

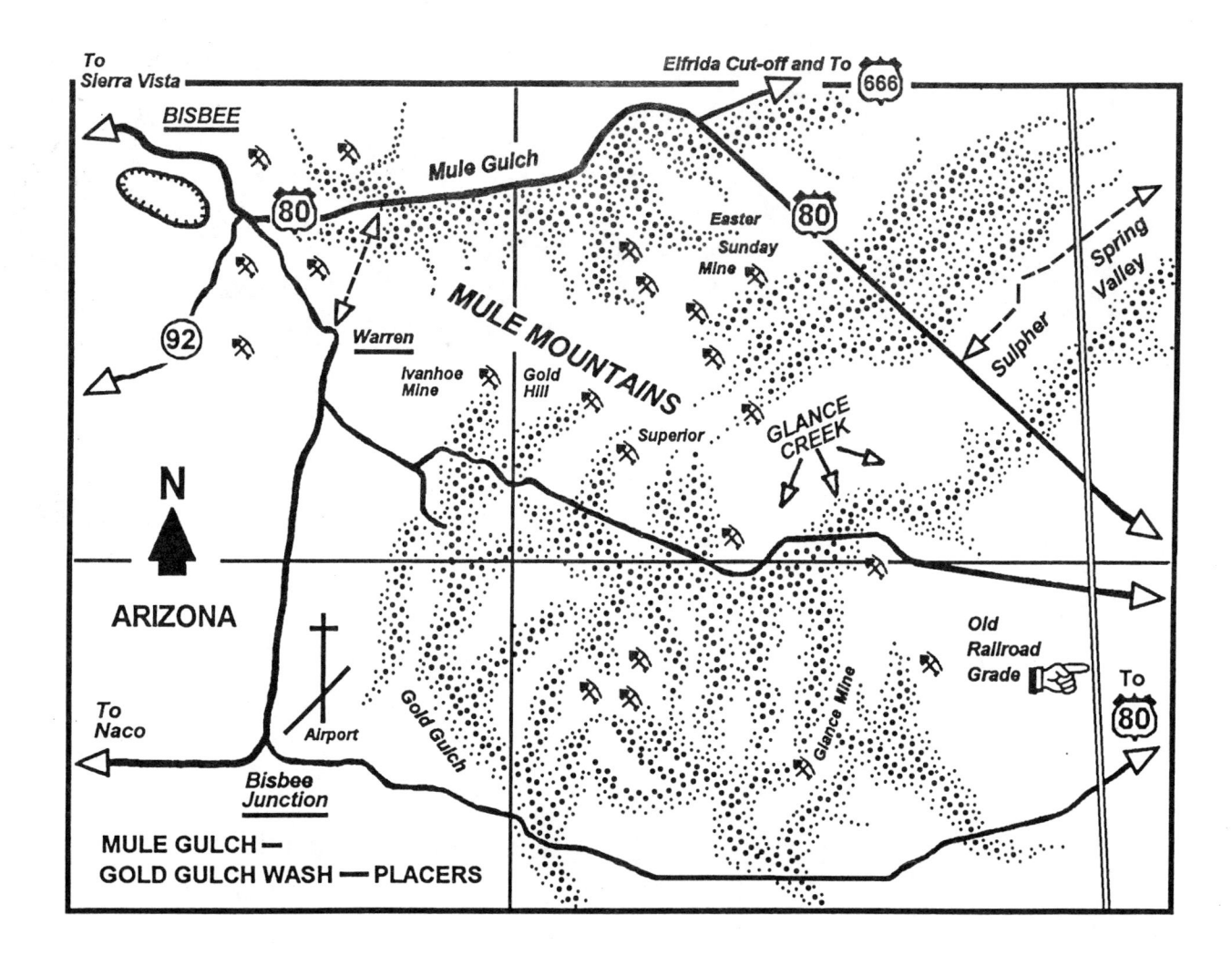

143